SYSTEMS MANAGEMENT TECHNIQUES FOR BUILDERS AND CONTRACTORS

SYSTEMS MANAGEMENT TECHNIQUES FOR BUILDERS AND CONTRACTORS

PAUL G. GILL

Director of Master Planning
The Ingalls Shipbuilding Division
of Litton Industries, Inc.

McGraw-Hill Book Company

New York   San Francisco   Toronto   London   Sydney

SYSTEMS MANAGEMENT TECHNIQUES FOR BUILDERS AND CONTRACTORS

Copyright © 1968 by McGraw-Hill, Inc. All Rights Reserved. Printed in the United States of America. No part of this publication may be reproduced, stored in a retrieval system, or transmitted, in any form or by any means, electronic, mechanical, photocopying, recording, or otherwise, without the prior written permission of the publisher. *Library of Congress Catalog Card Number* 68-20989

23236

1234567890 HDBP 7543210698

TO MAURA

PREFACE

During the past two decades there has been an unprecedented level of management challenge and innovation in American industry which has generated an impressive output of powerful new management techniques.

The quest of technology in the space age created explosive advances in the physical sciences which left in their wake a critical gap between management know-how and technological progress. To bridge this gap, a whole new generation of management tools was developed. One of the most fundamental and far-reaching of these management innovations was the *systems management approach,* an exciting new concept of an organization as a totally integrated system of operations.

During this same period, there also emerged a critical need for more dynamic management tools for organization development and operations control among firms in the home building industry to match the explosive increase in demand for housing. However, the full impact of the new generation of management techniques was not felt in this industry because the mainstream of management innovation was in those industries directly exposed to the military and space needs of the nation.

The purpose of this text is to provide builders and contractors with some practical and proven management guidelines for implementing the systems management approach in the development of their organizations and the control of their overall operations. It has been prepared specifically for use by the large number of firms at the grass-roots level of the construction industry, where the need for more powerful management techniques is most urgent. However, the text also provides basic standards of management against which larger firms may evaluate the effectiveness of their organization development and operations control methods.

Part I of the text outlines the fundamental principles of systems management planning, scheduling, and control as they apply to home building operations. The Gantt chart, critical path method, and systems management techniques are thoroughly reviewed and compared as to their advantages and limitations. The book then proceeds with a demonstration case to illustrate the application of the systems management approach for the planning, scheduling, and control of a fifty-semicustom-home construction program.

The demonstration takes the reader from the inception of the construction program through its many stages of actual construction operations: requirements planning, construction planning and scheduling, resource planning and scheduling, monitoring and controlling the flow of construction activities and resources, and the management of the entire construction program as one completely integrated system of operations.

Parts II and III outline the use of systems management techniques for developing and controlling home building organizations. The principles of organization structures, the delegation of duties and responsibilities to the management team, and the establishment of business systems and procedures necessary for day-to-day home building management are explained in detail. In addition to defining what management tools are essential for organization development and control in home building, the text demonstrates how these tools are actually implemented through organization charts, job descriptions, and systems and procedures which cover in great detail every major aspect of business administration in a home building organization.

The systems management techniques outlined herein were publicized during 1965 in the *House and Home* and the *Long Island Builder* magazines following their successful application to home building. As a result of this publicity, I received a large number of requests from home builders throughout the United States and Canada for a construction mangement text for use in their firms. With the encouragement of Mr. Ben Lindberg, one of the most esteemed management consultants in the industry and former associate professor at the Harvard Business School, I set to the task of preparing this book. During the course of its preparation, the systems management approach was successfully introduced to a number of medium- and large-sized shopping center and home building organizations.

My business experience has been focused on the application of the systems management approach for organization development and operation control purposes for firms in the defense, construction, and maritime industries, and my formal education includes a masters' degree in business administration from the Harvard Business School. The systems management techniques which

Preface

I present in this text are based on those used for weapons systems management by the Department of Defense and the functional management techniques in general use in other fields of management.

I wish to express my sincere appreciation to those builders and contractors who contributed in one way or another toward the preparation of this book. In particular, I am indebted to Fred Epstein and Norman Zaret, Development Corporation of Puerto Rico, San Juan, Puerto Rico; Joseph P. Lenny and Alfred V. Masullo, Eastern Colonies Corporation, Haddonfield, New Jersey; Saul Muchnick and Herman J. Rudy, Northwood Enterprises, Ltd., Huntington, New York; Joseph W. Proctor, Jr., Johnson and Proctor, Cold Spring Harbor, New York; John Frost, Old Hills Construction Corp., Greenlawn, New York; and John G. Walsh, Jr., Director, Business Management Department, National Association of Home Builders, Washington, D.C.

Paul G. Gill

CONTENTS

Preface   vii

PART I   OPERATIONS PLANNING, SCHEDULING, AND CONTROL   1

CHAPTER 1   OPERATIONS CONTROL TECHNIQUES   7
 Gantt Charts   7
 The Critical Path Method   9
 Systems Management Techniques   12
  Construction Systems Management   16
  Systems Management Matrix   18
 Program Control Charts   21

CHAPTER 2   DEVELOPING OPERATIONS PLANS   28
 Operations Planning   28
  Fundamentals of Planning   29
  Prerequisites for Planning   31
 Requirements Planning   33
  Construction Manpower   34
  Construction Materials   37
  Changes and Selections   38
  Permits and Inspections   39
 Construction Planning   40
  Construction-time Estimates   40

        Construction-flow Plan   44
        Resource Lead Time   47
        Resource-flow Plan   55

CHAPTER 3   ESTABLISHING OPERATIONS SCHEDULES   58
    Operations Scheduling   58
        Custom Flow   59
        Production Flow   59
    Construction Scheduling   60
        Job Schedules   61
        Production Schedules   62
    Resource Scheduling   64
        Subcontractor Trades   64
        Plant Labor   65
        Purchase Orders   67
        Changes and Selections   67
        Job Inspections   67

CHAPTER 4   IMPLEMENTING OPERATIONS CONTROLS   68
    Production Control Techniques   68
        Construction Status Evaluation   69
        Construction Variance Control   74
    Program Control Techniques   77
        Program Status Evaluation   80
        Program Variance Control   84

PART II   ORGANIZATION DEVELOPMENT AND CONTROL   87

CHAPTER 5   ORGANIZATION FOR GROWTH   90
    Functional Structures   90
    Economic Characteristics   95
    Format for Structural Growth   97

CHAPTER 6   DELEGATION AND CONTROL   100
    Job Descriptions   100
    President   102
    General Manager   102
    Project Engineer   104
    Finance Manager   105
    Construction Manager   106
    Marketing Manager   108
    Office Secretary   110

PART III   SYSTEMS AND PROCEDURES   111

CHAPTER 7   ENGINEERING PROCEDURES   115
    Site Analysis Surveys   115

Contents xiii

                Project Plans and Specifications   117
                Construction Change Orders   118
                Change Order Control   122

CHAPTER 8   FINANCE PROCEDURES   124
                Construction Loan Applications   124
                Construction Loan Closings   125
                Construction Loan Receipts   126
                FHA Conditional Commitment Applications   128
                FHA Firm Commitment Applications   129
                Title Closings   131
                Escrow Agreements   132

CHAPTER 9   CONSTRUCTION PROCEDURES   134
                Building Permit Applications   134
                Site Preparation   135
                Daily Field Reports   136
                Daily Field Report Register   138
                Job Status Record   140
                Construction Status Reports   143
                Construction Schedule Variance Control Reports   145
                Customer Complaints   147
                Customer Complaint Control   149
                Job Work Orders   151
                Certificate of Occupancy Applications   154

CHAPTER 10   PURCHASING PROCEDURES   155
                Requests for Quotation   155
                Purchase Orders   158
                Purchase Order Register   160
                Vendor Certifications   162
                Vendor Register   163

CHAPTER 11   SUBCONTRACTING PROCEDURES   165
                Subcontractor Agreements   165
                Subcontractor Work Schedules   167
                Subcontractor Certifications   170
                Subcontractor Register   171

CHAPTER 12   INSPECTION PROCEDURES   173
                Town Building Department Inspections   173
                Town Health Department Inspections   174
                FHA Compliance Inspections   177
                Lender Compliance Inspections   178
                Customer Acceptance Inspections   180
                Construction Inspection Record   183

CHAPTER 13   MARKETING PROCEDURES   186
                Sales Option Agreements   186
                Preliminary Sales Agreements   187

Contracts of Sale 188
Customer Service Register 189
Customer Selections Reports 192
Weekly Sales Reports 194

CHAPTER 14 FILING PROCEDURES 196
Project File System 196
Job File System 199

Bibliography 201

Index 205

Part I

OPERATIONS PLANNING, SCHEDULING, AND CONTROL

Since World War II a whole new generation of sophisticated operations control techniques have been developed and implemented to solve the complex business problems confronted in the space age. Without their innovation, many of the scientific accomplishments realized in America's leading industries over the past two decades would never have been possible. Their impact on the field of management science has been as profound as the technological breakthroughs achieved in the field of physical science. Their tremendous success has escalated learning in the management sciences toward the threshold of revolutionary new techniques and systems for controlling entire business organizations.

The interest of management scientists in leading universities has been spurred toward intense research and development to further expand the applications of these new operations control techniques. Their goal is to construct more powerful management control tools that totally integrate the entire operations system of business organizations. This would permit the complete unification of all operating elements in a business organization into a totally mechanized system of operations. Through the use of operations research and electronic data processing techniques, it will soon be possible to similate entire business organizations mathematically so that management may predict in advance the probable effect of all major decisions before action is taken. This goal will be reached by the management scientists in the very near future.

The development of these sophisticated new operations control techniques for management may be attributed largely to the growth needs of manufacturing firms in the wake of World War II. The phenomenal increase in the demand for consumer and industrial goods, in parallel with the expanded government needs for complex military and space requirements, exerted an unprecedented pressure for production output by manufacturers. The conventional Gantt-Taylor operations control techniques used by management were quickly outmoded. The need for manufacturers to improve their products, increase production output, expand facilities, and modernize operations processes was most critical. The situation demanded fresh management thinking and new operations control tools to solve complex business problems which had never before existed. Operations research and computer programming techniques provided the much-needed management solution to operations control for manufacturing firms. These new management techniques not only provided more effective operations control over their operations systems, but also facilitated the rapid growth of their organization structures to satisfy their market demands and increase their profit margins.

During this same period, the home building industry was also under great pressure to expand to satisfy the explosive increase in demand for postwar housing. However, the expansion problems confronted by the home building industry were vastly different from those faced by manufacturing. The need for increased industrial output in the manufacturing field was accomplished largely through the full-scale expansion of plant facilities and equipment modernization by existing companies. The massive need for consumer housing could only be satisfied through the rapid influx of a large number of new home building firms into the industry.

The diverse routes to expansion followed by the manufacturing and home building industries led to their developing extremely different economic structures. Manufacturing industries grew pyramid-wise, primarily because of the expanded outputs of existing large firms which dominated the production outputs in their fields. On the other hand, the home building industry expanded plateau-like through the entry of a large number of small new construction firms, only a small portion of which constructed more than thirty homes a year.

Operations planning, scheduling, and control are fundamental management problems which are common to both the home building and manufacturing industries. However, there is a striking difference between the operations environments of the firms in these industries and the operation control problems with which they are confronted.

Manufacturing firms maintain their plants and production facilities in fixed geographic locations. To increase their production volume or change their product line in response to market demand merely requires additional increments of labor and material inputs, or modifications in their existing production lines. If a home builder wants to expand his construction operations and volume, it is necessary that he acquire and develop construction sites. Manufacturers can remain fixed in their geographic locations and expand their production volumes through the addition of more men, material, and equipment to their existing production lines, while home builders must relocate their operations sites and improve more land to increase their construction volumes. Operations controls can be formulated and implemented by manufacturing firms to remain fixed for long periods of time. They need only be modified to reflect improvements in control methods or changes in the production process. Home builders, however, must formulate and introduce new operations control systems for each new construction program.

Operations planning in manufacturing firms is based on their long-range growth goals. Their long-range plans are subdivided into current, short-term, and intermediate-range plans, which may span a ten-year period or longer. The plans are interconnected to guide the firm along each phase of its long-range planning path to ensure that the ultimate goals of the organization are fully realized. Because of the fixed nature of production operations in manufacturing firms, the establishment of long-range planning objectives is both feasible and logical. In comparison, operations planning in the home building field is extremely short-run in nature and primarily on a project basis because of the need for home building firms to continuously relocate with shifts in their market demand. It is not possible for home builders to forecast their plans from one project through a succession of projects and thereby interconnect each of these plans into a long-range planning goal.

The trend in manufacturing toward automated and semi-automated plant operations has accelerated since the development of computer control systems in the mid-fifties. As a result, management in this field has been focusing an increasing amount of attention on machine performance because of its high degree of reliability and predictability. Since there have been no major technological advances in the mechanized construction of homes, builders find it essential that they maintain firm supervision and control over the performance of craftsmen and labor in their construction operations. Construction management, therefore, requires much tighter coordination and integration of human resources than does manufacturing to accomplish its operational objectives.

Inspection requirements imposed by federal government agencies on manufacturing firms engaged in defense production are not nearly as burdensome as those placed on home builders by combined local, state, and federal building regulations. Federal regulations established by each of the military and space procurement agencies are fairly uniform. For advisory and inspection purposes, representatives of the federal agencies are usually located in manufacturing plants that are under government contract. But there is no uniform code of regulations to govern quality standards and inspection requirements for home builders. Building regulations vary from town to town, county to county, state to state, and from one federal agency to another. As a result, home builders are subjected to conflicting quality standards and inspection requirements from every quarter of government. The continuous liaison required with local planning boards, building inspection departments, health and water inspections departments, etc., exerts a most unfavorable influence on operations control and growth for home building organizations.

The physical aspects of home building operations are gargantuan in comparison with those of the average manufacturing firm. Taking into consideration the development of land and the concurrent construction of a large number of homes at the same time in one project, home building operations can become extremely complex. Management's ability to control construction operations must be commensurate with the size of the construction program. The increased complexity in the management of large construction projects seriously limits the ability of the average builder to expand his operations. Possibly the most serious restraint imposed on the growth of home builders' organizations has been the lack of operations control systems designed specifically for their needs. The management techniques presently used by builders were originally developed for the solution of operations control problems confronted by companies in other industries. Their application to the management control problems of builders has met with limited success because of the distinct difference in character between home building and manufacturing operations.

What complicates the role of management in home building organizations is the fact that the average builder must be both a functional and systems manager in order to successfully expand his business. The functional management approach provides the means for the planning, scheduling, control, and evaluation of all functional departments of the organization. The systems management approach requires the ability to effectively integrate all administrative and construction activities into a total operations system. Functional management demands a thorough knowledge of the entire construction business process, while systems management calls for the ability to smoothly integrate the flow of all administrative activities into the mainstream of the firm's construction operations.

This dual functional/systems management capability is a basic requirement for all builders, regardless of their construction volume. There must be a continuous balance in the focus of management on both the functional and systems aspects of overall operations. In no other industry is such a fine balance of management talent so critical to organization growth. The operations control tools required by home building management must provide for both the functional and systems control of total operations. The management tools that have been transplanted into the home building industry from other fields have been of value in specific areas, but have not provided a total operations control technique for the management of overall construction operations.

The continuous flow of new management concepts and techniques throughout manufacturing industries during the past twenty years has contributed substantially to both their operation control and organization growth. This cross-pollination of advanced management ideas and methods in manufacturing fields does not occur in the home building industry. Many of these innovations in management methods originated in manufacturing firms engaged in military and space programs. Under continuous government pressure for increased productivity and technological improvements in the weapons and space systems that they produce, these firms meet their program objectives by responding to a constant demand for fresh management approaches to operations control problems.

Innovation in management, as in technology, is the product of an intellectual or scientific environment which is highly conducive to the inception of fresh concepts, their development for experimentation, and their implementation for productive returns. Manufacturing firms cultivate innovations in their management thinking through academic and financial incentive systems, continuous acquisition of progressive management talent, and participation in advanced management "think" seminars sponsored by leading universities and professional management societies.

The operations environment of the average home builders' organization does not provide the intellectual or scientific stimulus required for the development of fresh management concepts. The limited capital structure of the average builder does not permit the extension of grants to employees in academic or financial incentive systems, their participation in sophisticated management seminars, or their acquisition of professional staff talent to aid in the solution of their management problems. As a result, construction management has had to look to other industries for new tools and techniques to solve their particular operations control problems.

Presented in Part I of the book are the fundamentals of operations planning, scheduling, and control for home building operations, and also a practical step-by-step demonstration of the use of some extremely effective operations control techniques for home builders. The techniques presented have been developed in the field for specific use on home builders' construction programs and incorporate the basic principles of:

1. The Gantt chart
2. The critical path method
3. Program evaluation and review techniques
4. Weapons system management concepts
5. Operations research
6. Computer programming techniques

While the operations control techniques illustrated are sophisticated in principle, they are demonstrated in a practical manner to facilitate their grasp by the reader.

CHAPTER 1

Operations Control Techniques

GANTT CHARTS

The Gantt chart was developed during World War I by Henry L. Gantt, one of the pioneers in the scientific management movement during that period. It was designed for use in the planning and scheduling of production operations and was immediately accepted as a revolutionary improvement over existing management control methods. Because of its functional simplicity, versatility of application, and visual effectiveness for control purposes, it was widely adopted as a fundamental tool of management. Most commonly known as a *bar chart*, its primary application in manufacturing has been for planning, scheduling, and controlling production operations. It provides management with a practical means of coordinating in graphic form the work to be performed and the time required for its accomplishment. For production planning, the work activities to be performed are listed vertically on the left-hand scale of the chart (see Fig. 1.1) and the time constant is scaled horizontally across its top. The time scale may be expressed in any suitable divisions and subdivisions—months, weeks, days, or hours.

The bar chart is an indispensable management tool in manufacturing firms, where operations are highly departmentalized from standard parts production through their component, assembly, and fabrication stages of operations. It is used to coordinate and control the flow of parts, components, subassemblies, and assemblies into finished end products. Each separate stage of operations is monitored with the use of planning, progress, load, and lead-time bar charts. Collectively, they provide manufacturing management with the means necessary to integrate all departmental stages of operations into a total scheme of operations. In this manner, management is assured that each facet of operations is dove-

No. Unit	Gantt Chart Production Schedule														
	Work Weeks														
	1	2	3	4	5	6	7	8	9	10	11	12	13	14	15
2501															
2502															
2503															
2504															
2505															
2506															

Figure 1.1

tailed into a master manufacturing plan to accomplish overall operations objectives.

Although originally developed as a manufacturing control technique, bar charts have been widely used by home building firms for the planning and scheduling of construction operations. However, their application in home building differs from their fragmental use in manufacturing, where separate bar charts are used for each major stage of operations. In home building they are used as a master control device for planning, scheduling, and controlling all phases of construction. The bar chart has been broadly accepted as a fundamental management control tool throughout the home building industry.

During recent years the critical path method (CPM) of project planning and scheduling developed for use in the heavy construction industry has been applied by firms in the home building industry. While this project management technique has not won widespread acceptance by home building firms, its applications have brought to light some serious shortcomings in the bar chart technique. The limitations uncovered in the use of bar charts for home building operations are as follows:

1. Planning: They do not show the interrelationships between job activities that are critical and those that may be performed concurrently

with others; they are either oversimplified in one area of operations or too detailed in another; they do not show a logical sequence of activities to be performed; they do not show the "slack" that may exist for noncritical activities; they do not show the critical path of operations, and as a result all activities receive an equal amount of attention; they permit the examination of activity problems only when they occur; and they fail to show a detailed, integrated plan of operations.
2. Scheduling: They are developed on a time basis alone and cannot be used to advise management what resources are available to speed up a critical activity; they cannot predict a correct course of action when delays are imminent; they are not adequately detailed to spell out tradeoffs between time and effort; they do not effectively integrate the time of performance of activities from one job to another; and they cannot predict where and how to expedite operations.
3. Control: Bar charts require a great deal of study to determine where jobs actually stand in relation to each other, and they do not provide an effective means of management control over all activities to be performed and the resources required for their performance.

Although they do have many limitations when compared with the critical path method, they also remain the most widely accepted management control tool in home building because of their simplicity and visual control value. They cannot schematically integrate all activities in a construction plan, cannot interrelate all activity and resource schedules, and cannot provide management with integrated activity and resource controls as the critical path method can. However, to date their major advantages, simplicity and visual control value, have outweighed their apparent limitations, and at present bar chart usage in home building far exceeds applications of the critical path method.

THE CRITICAL PATH METHOD

One of the powerful new management control tools developed during recent years is the critical path method (CPM). E. I. du Pont de Nemours first conceived of the technique in 1956 when it initiated a study program to develop a computer method of planning and scheduling for plant maintenance and construction programs. In 1957 Univac Applications Research Section joined the program, the result of which was CPM, a technique which presented project logic in graphic form. It is represented by an arrow diagram or network which shows the interrelationship of one job (activity) in the project to the whole (see Fig. 1.2).

It was specifically designed for the management and control of large, complex, one-of-a-kind projects which involved thousands of job activities and spanned long periods of time.

It was accepted almost immediately in the heavy construction industry. At present between 15 and 20 percent of the heavy construction work now in progress is CPM planned and scheduled, and its use is expected to further increase over the next few years. It has been used for project management in the construction of industrial plants, bridges, hospitals, schools, apartment buildings, laboratories, and truck terminals. In other

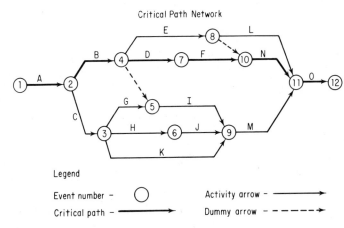

Figure 1.2

fields its applications include financial closings, road maintenance, market planning, and factory maintenance. Its potential use for managing and controlling complex projects is unlimited.

The critical path method enables management to see the significant interrelationships between all tasks to be performed on a project. It provides a means for highlighting exactly where trouble spots are likely to occur. While it cannot tell management what decisions it must make, it supplies important information pertinent to decision making and serves as a guide to assess the effect of alternate decisions. It serves as a management tool for defining and integrating events which must be accomplished on a timely basis to assure completion of the project objectives on schedule. It defines areas of effort where tradeoffs in time, resources, or performance will enable management to meet scheduled dates. The most distinguishing characteristics of CPM are:

1. It enables management to plan the best possible use of resources to achieve a given goal within overall time and cost limitations.

2. It allows executives to manage complex one-of-a-kind projects as opposed to repetitive production-like operations.
3. It helps management handle the uncertainties involved in programs where no standard time data of the Taylor-Gantt type are available.
4. It utilizes a *time-network analysis* as a basic method of approach to determine manpower, material, and capital requirements.

The critical path method is the most effective project management tool yet conceived for the control of heavy construction projects, unmatched by any other management control technique. It is capable of molding vast, cumbersome projects into totally integrated and manageable systems of operations. It represents a major breakthrough in management know-how in the heavy construction industry.

The critical path method has met with limited acceptance in the home building field. Though its application has proven successful for a number of home building firms, it has not won the rapid and widespread approval that it has gained in the heavy construction field. The reasons for home builders, in general, not adopting the critical path method are fundamental, as follows:

1. The critical path method was not developed for use on repetitive-type production operations, such as those in the home building field. It was designed for planning, scheduling, and controlling large, one-of-a-kind construction projects such as industrial plants, bridges, warehouses, and similar structures. Home building is basically a production-oriented operation in which a number of housing units are constructed concurrently and a large degree of duplication is present in the production process for each unit. The development of separate CPM controls for each housing unit, as would be required for each heavy construction project, would be impractical in home building and would restrict rather than enhance the flow of production operations.
2. In comparison with the bar chart, the critical path method is a complex management control technique. The bar chart is simple to prepare, and its two coordinates, time and effort, are simple to visualize on the scales of the chart. The diagram used for the critical path method is basically nondimensional and does not provide scales for the measurement of time and effort. Instead, it provides arrows to represent activities to be performed, and their length is not proportionate to the time required for their performance. The configuration of the CPM diagram is not geometric and does not lend itself to the measurement of time and effort in relation to the size of its activity arrows. The configuration of the diagram is in logic network form, and its principal functions are to define the sequence in which all

activities are to be performed, their interrelationships and interdependencies, and, after each activity has been time-estimated, to determine the longest and most critical activity path in the network. The critical path method is a much more complicated management control technique for home builders to apply than the bar chart.

3. Since the critical path method is essentially a computer-oriented technique, its application costs are high. In fact, the cost savings which may be realized with the use of the critical path method may be more than offset by the cost of computer services if the size of the project is not large enough to warrant its application. To illustrate, the estimated break-even point for the application of computer-CPM to home building operations is a thirty-home construction project. The computer service cost per home may average $150 a unit if the homes are constructed concurrently (much more if they are not). If the number of homes to be constructed is below this estimated break-even volume, then there are no cost savings to be realized from the computer-CPM application. If the home building volume is above this level, then the cost savings possible must be weighed against other tangible and intangible administrative costs involved in its application.

4. Computer-CPM requires the preparation, processing, and distribution of voluminous operations data. The schedule status, monitor reports, and other information furnished by the computer are extremely detailed and lengthy, requiring continuous analysis and revision by management. The dilemma presented by computer-CPM is that management is apt to devote more time evaluating computer data in the office and considerably less time to quality control in the field, to the detriment of the finished products.

The critical path method as designed for use in heavy construction has not provided any panacea for the operations control problems confronted by home builders. While it provides an extremely effective solution for integrating the overall operations of a construction program, its advantages have been greatly outweighed by those of the bar chart in home building because of its complexity.

SYSTEMS MANAGEMENT TECHNIQUES

Over the past two decades there has been a technological explosion in military weapons systems concepts which has spawned a tremendous need for new management techniques. During the early 1950s it was taking as many as fifteen years for the American armed services to

Operations Control Techniques

bring one weapons system (ballistic missile, strategic bomber, tactical fighter plane, etc.) through the design and development stages into operational status. The Soviet Union was accomplishing the same in half the time. The need for fresh management approaches to the development and control of America's complex weapons systems programs was critical.

The first major breakthrough in the management of these complex programs was made by the U.S. Air Force when it conceived of the "systems management concept" and established weapons systems project offices for their management and control. Its implementation involved the installation of an additional layer of management and authority over and above that in the existing functional organization structure (see Fig. 1.3). Systems management procedures were established which cut across the traditional functional lines of organization to ensure that high-priority weapons systems received continuous attention beyond the specialized interests of any one particular department of the organization. At the headquarters level, systems staff offices were established and assigned the responsibility for weapons systems management with the aim of integrating and coordinating the staff activities of all functional departments to ensure that weapons systems management plans, schedules, and controls were effectively implemented. At the field level of operations, systems program offices were installed for the purpose of integrating and coordinating the systems-oriented activities of all functional field offices to ensure that all weapons systems development and operational goals were realized.

The impact of the systems management approach on the armed services weapons systems programs was revolutionary. Through its successful implementation the critical void between management control know-how and weapons systems technology was rapidly closed. The lag between American and Soviet weapons systems design, development, and deployment was rapidly eliminated. The systems management approach made it possible for the armed services to totally integrate and control their functional organization structures in order that their weapons systems program objectives could be achieved as planned.

The second major management breakthrough realized by the armed services during the 1950s was the development of the program evaluation and review technique (PERT). Similar to the critical path method used in the heavy construction industry (see Fig. 1.4), it is based on the logic network system for planning, scheduling, and controlling large, complex weapons systems programs. It was conceived by the Special Projects Office of the U.S. Navy for use in the planning and control of time schedules in the Polaris Fleet Ballistic Missile Program. As a result of its outstanding success on this program, PERT was rapidly

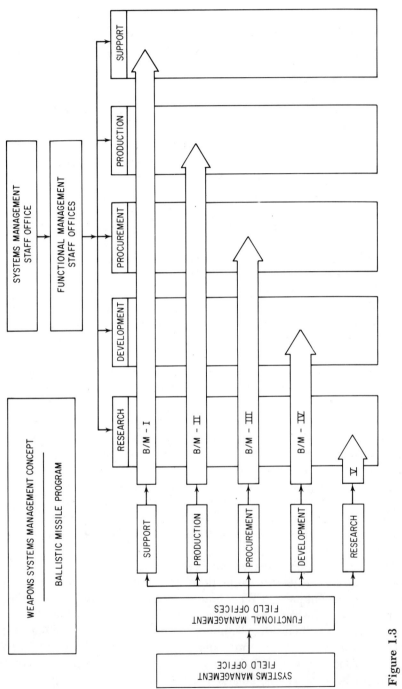

Figure 1.3

adopted throughout the armed services for the management and control of complex weapons systems development programs.

PERT represented another significant step forward by military management in the development of totally integrated management systems for controlling complex weapons systems programs. The improved planning that it provided offered a sound basis for scheduling weapons systems development and the means by which their status could be measured and problem areas isolated to permit corrective action. In addition,

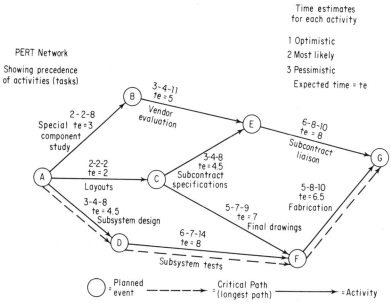

Figure 1.4

it provided managers at all levels with the integrative discipline necessary for the successful attainment of the prime and supporting objectives of weapons systems programs.

While the systems management concept furnished the management approach necessary for the total integration of functional and systems operations flows in military organizations, the logic network technique made it possible to completely coordinate and control all activities involved in the development of complex weapons systems (Fig. 1.5). It provided management with a logical approach to the design, development, and control of weapons systems programs, from the conceptual through the operational stages. As a result, weapons systems managers were able to allocate and control all functional activities and resources in a highly systematic manner to best meet their weapons systems development requirements. In addition, it made it possible for management to multiprogram and control a number of similarly complex weapons systems under concurrent development. The logic network is the most powerful management tool ever conceived for the development and control of complex weapons systems programs.

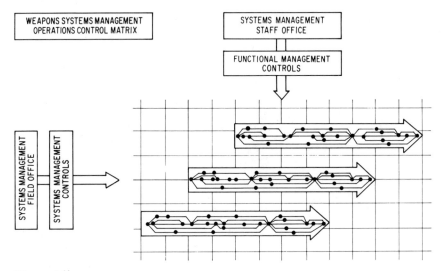

Figure 1.5

Construction Systems Management

The explosive growth in the nation's population during the past twenty years has generated an overwhelming need for better management methods in the home building industry. Home builders' expansion efforts have been severely restricted by the lack of effective construction management techniques to resolve their complex operations planning, scheduling, and control problems. The home building cycle, from the acquisition of raw land to the completion of construction on a housing unit, is extremely long when compared with that for other consumer products. The number of construction activities involved in the development of a housing community may run into many thousands, making it extremely difficult to coordinate and control the flow of overall operations. In addition, the large-scale production of housing is most difficult to manage because the resource requirements cut across many areas of industry and agencies of government.

Although conceived for the development and control of weapons systems programs, the systems management approach has also been successfully used for the coordination and control of home building programs.

Operations Control Techniques

The operating characteristics of home building organizations are similar to those of weapons systems organizations. Because of the complexities of their end products and their long lead-time cycles, both types of organization require functional and systems management controls for their management. Their organizations require a continuous balance of management control over the functional flows of their activities and their resources into the process of operations and over the systems flow of all housing units/weapons systems through their operations cycles.

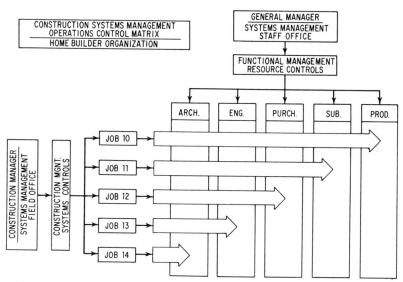

Figure 1.6

The balance of management control over functional and systems operations in their organizations can only be maintained through a division of management responsibility between their staff and field offices. Through this dual process of management control, both the functional and systems flows of operations through their organizations are interlocked into totally integrated operations systems.

The systems management approach in home building organizations is illustrated in Fig. 1.6. At the staff level of organization, the general manager has the overall systems management responsibility for construction operations, coordinating and controlling the functional flows of organization activities and resources required into the operations process, and ensuring that overall construction plans and schedules are carried out as programmed. At the field level of operations, the construction manager has the responsibility for coordinating and controlling the systems flow of all housing units through their operations cycles. Field management's responsibility is to ensure that there is a continuous flow of construction labor, materials, and other needs into the systems flow of housing units as they are processed through operations in accordance with established operations plans and schedules.

The successful introduction of the systems management approach in a home building organization requires the definition and delegation of duties and responsibilities to each member of management, and the development and establishment of sound systems and procedures for its implementation. These requirements are discussed and illustrated in detail in the sections "Delegation and Control" and "Systems and Procedures" which appear later in the text.

Systems Management Matrix

The original application for which logic networks (CPM and PERT) were developed was for the management and control of complex, one-of-a-kind weapons systems and heavy construction programs. In order to manage and control a number of similar programs, as shown for weapons systems in Fig. 1.5, separate networks had to be prepared and implemented for each. The networks for these complex programs had to be intricately interfaced with each other so that the flow of activities and resources required for the development of each could be logically interwoven from one program to the next.

The computational demands imposed on military management for the multiprogramming of large weapons systems programs were beyond the control capacities of conventional manual and mechanical calculation techniques. The extremely large number of activity and resource input and output calculations required to network one major weapons systems program alone was staggering.

Management's only feasible solution to the multiprogramming problem was to computerize the logic network data of each program with electronic data processing equipment. This made it possible to rapidly convert raw logic network data into urgently needed plans, schedules, and reports so that management could control a number of similar complex programs at the same time. In addition to interfacing the concurrent flow of each weapons program, the computer also interrelated the flow of all activities and resources between them.

The use of electronic computers for multiprogramming complex weapons systems programs offers significant advantages over conventional management methods, as it provides great computational speed and accuracy for processing large masses of data. However, the use of computers for home building operations is not so advantageous, primarily because the large volume of output data generated necessitates lengthy and thorough study by management to assess the overall status of the construction program. To illustrate, a construction program with fifty homes concurrently in production would, at minimum, require monthly computer reports for the evaluation of construction progress. The computer would print out a 3-page schedule report for each home under construction, or the equivalent of a 150-page schedule status document for the entire program. In addition, the computer would provide resource and subcontractor printout status reports, all of which management would find most difficult to mentally assimilate, evaluate, and then act upon before the next month's computer reports were available.

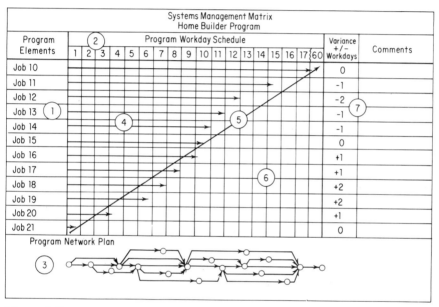

Figure 1.7

There is a simpler approach to operations data processing for home builders with the *systems management matrix* shown in Fig. 1.7. It is a visual management control technique which greatly facilitates the planning, scheduling, and control functions associated with multiprogramming the concurrent construction of a large number of housing units. The systems management matrix simulates the data processing characteristics of a computer system in that it permits rapid program data readin, has an immense data storage capacity, contains a logic network for programming all operations to be performed on each program element, and provides instantaneous readout data on the status of all operations on each program element. In effect, the systems management matrix technique "closes the loop" between the Gantt chart, critical path method, and computer programming as it provides management with a simple tool to graphically visualize, logically evaluate, and rapidly interpret the detailed and overall status of complex business programs.

A detailed description of each of the major characteristics of the systems management matrix as illustrated in Fig. 1.7 follows:

1. Program elements: lists vertically the jobs (housing units) to be constructed and their sequence of priority.
2. Program workdays: lists horizontally the total number of construction workdays required for the construction of the most complex housing unit in the program.
3. Program network plan: a schematic diagram (logic network) which illustrates and integrates all the activities to be performed, their sequence of operations, and the time required for their individual and overall accomplishment in the construction of the most complex housing unit in the program.
4. Program element progress bars: horizontal bars drawn over the program element (job) schedule dates to indicate the actual progress realized to date on the program element as related to the program network plan.
5. Program progress curve: a diagonally sloped regression line which moves from the left to the right of the matrix as performance is posted on each program element progress bar. The program element progress curve is actually an imaginary line, the configuration of which is determined by the progress shown by the bars for each program element. If the progress curve has a straight-line movement across the matrix, it reflects balanced management focus and construction progress on each job in the program. If, on the other hand, the progress curve assumes the shape of an S curve or some other configuration, it visually signals that the overall construction program is out of balance.
6. Program schedules: provide the schedule dates for the construction of each housing unit from their scheduled start to completion dates (horizontal readout). The program schedule also indicates the scheduled start and completion dates for subcontractor trades, inspection agencies, etc., on each job in chronological sequence (vertical readout).
7. Program variance control: provides management with the means to measure and evaluate performance and progress on each housing unit under construction. The variance of construction performance from program schedules may be measured for each job (in construction workdays) to determine whether critical and noncritical work activities are ahead, behind, or on schedule. The remarks column

makes it possible to note and flag critical problem areas in the program as they develop and facilitates the practice of management by exception.

The systems management matrix may also be considered an operations input-output model which graphically simulates the flow of all construction activities, resources, and jobs through the home builders' organization. The model provides a schematic means for integrating the flow of all construction activities and resources into the operation process with the flow of all construction jobs through the operations system. The operations inputs in the model are the vertical flows of activities and resources generated by each functional department of the organization. The operations outputs are the horizontal flows of jobs through the operations process as they undergo their various stages of construction and are ultimately phased out of the operations system as completed housing units.

PROGRAM CONTROL CHARTS

The systems management matrix provides home builders with the basic format required for the development of a simple yet effective graphic management tool with which to plan, schedule, and control their construction operations. The management control tool developed for this purpose is the *program control chart* shown in Exhibit I, which has been successfully used for operations control on a wide variety and volume range of residential housing construction programs.

For flexible planning, scheduling, and control purposes, the program control chart is made up of two parts, a program planning sheet (matrix) and a program control sheet (transparent overlay). Construction plans and schedules are plotted in pencil (to simplify revisions) on the program planning sheet, and construction performance, variance, and status are posted with colored crayons (to simplify updating) on the program control sheet. A detailed description of the information entered on each of these sheets as illustrated in Exhibit I follows.

1. Program Planning Sheet:
    a. Job No.—left-hand vertical column
    b. Construction Work Days—uppermost horizontal column
    c. Construction Schedules—horizontal schedule columns for each job
    d. Construction Plan—CPM network in center of sheet
    e. Construction Resources—purchasing, subcontracting, plant labor, and inspection requirements shown at bottom of sheet
    f. Construction Activity Starts—right-hand center of sheet

2. Program Control Sheet:
   a. Construction Progress Bars—drawn horizontally over each job schedule to show construction progress made
      (1) Critical Path Progress—top progress bar drawn with red crayon to show critical path progress on each job as related to critical path (centerline) on Construction Plan
      (2) Concurrent Path Progress—bottom progress bar drawn with black crayon to show concurrent path progress on each job as related to concurrent paths on the Construction Plan
   b. Job Status and Date—date posted and construction status entries made for each job
      (1) Critical Activity Status—critical construction stage each job is in on the date of posting, entered with red crayon
      (2) Concurrent Activity Status—concurrent construction stage each job is in on the date of posting, entered with black crayon
   c. Work Variance—number of work days each job varies from construction schedule on the date of posting
      (1) Critical Activity Variance—the number of days that each job is ahead (+) or behind (−) schedule on its critical work path on the date of posting, entered with red crayon
      (2) Concurrent Activity Variance—the number of days that each job is ahead (+) or behind (−) schedule on its concurrent work path on the date of posting, entered with black crayon

The program control chart provides management with a tremendous capability for the manual processing, storage, and control of construction operations information. To illustrate this information-handling capability, let us assume that the fifty-job home-construction program mentioned earlier for computerization is multiprogrammed on the program control chart (see Exhibit I). Each housing unit has over 400 information bits associated with its actual construction—that is, start, performance and completion dates for customer selections, construction changes, material ordering, subcontractor work, government inspections, etc. Once the program network plan (shown in the center of the chart) is developed for the program, it requires but several minutes for the home builder to manually post the scheduled operations dates for the complete construction of the job. This manual posting automatically provides readout data for the 400 information bits required by the home builder to complete the job. A program control chart for a fifty-job construction program provides instant readout capability on over 20,000 bits of information which will ultimately be required by management for its completion. With the use of an electronic computer, it would require a 150-page schedule report to provide the same schedule information which is furnished on the one program control chart.

Operations Control Techniques

The immense amount of detailed construction operations information made available to home builders with the use of the program control chart is illustrated in the following outline.

1. Construction—provides management with a firm control over total construction operations by indicating:
   a. Every work activity to be performed for the construction of each house (job)
   b. Those construction activities which make up the critical path on each job and govern the length of time required to complete the total job
   c. Those construction activities which make up the noncritical paths on each job and may be performed concurrently with critical work activities and need not hold up completion of the job as scheduled if they are delayed
   d. The scheduled start and completion dates for work activities on each job
   e. The time required to perform work activities on each job
   f. The interrelationship between work activities on each job and between jobs
   g. The actual construction progress realized to date on each job under construction
   h. The number of workdays ahead or behind scheduled performance on each job under construction
   i. The number of workdays ahead or behind scheduled performance on critical activity paths for each job
   j. The number of workdays ahead or behind scheduled performance on noncritical paths for each job
   k. The current status of construction progress on each job
   l. The number of workdays expended to date for the construction of each job
   m. The number of workdays remaining to complete the construction of each job
   n. The number of workdays required to construct each house
   o. The plan of operations for the concurrent construction of all jobs in the project
   p. The scheduled calendar starts, construction, and completion dates for each job

2. Subcontractors—provides management with an effective subcontractor control system by:
   a. Showing the schematic interrelationship of all subcontractor trade activities on each job
   b. Showing the progressive work flow of subcontractor trades from one job to the next

COMPARATIVE ANALYSIS
OPERATIONS CONTROL TECHNIQUES FOR HOME BUILDERS

Factor Comparison	Gantt Chart	Critical Path Method	Program Control Chart
Installation			
1. Training of personnel to operate and maintain	No problem. Only one person required.	Requires extensive training of two or more qualified personnel.	No problem. Brief training or study by one person required.
2. Orientation of personnel concerned	No problem.	Extensive orientation required.	Only brief orientation required.
3. Records system required	No problem. Majority of records maintained at working level.	Extensive and complicated.	No problem. Operations data recorded on chart. Other records maintained at working level.
4. Special requirements	No outside consulting services required.	Requires outside consulting services.	No outside consulting services required.
Operation			
1. Updating	No problem. Maintained on daily basis. Updated by construction superintendent.	Considerable input data required. Processed by computer.	No problem. Maintained on daily basis by construction superintendent.
2. Monitoring	Good. Visual control.	Fair. Computer reports control.	Very good. Visual control.
3. Need for computer	Not required.	Required.	Not required.
4. Outputs	Graphic display. Readily analyzed by visual inspection.	Computer tabulates runs. Requires group presentation. Large data output.	Graphic display. Readily analyzed by visual inspection.

Figure 1.8

COMPARATIVE ANALYSIS
OPERATIONS CONTROL TECHNIQUES FOR HOME BUILDERS
(*Continued*)

Factor Comparison	Gantt Chart	Critical Path Method	Program Control Chart
Planning and Scheduling			
1. Construction	Poor. Does not show activity interrelationships. Possibility of omissions great.	Good. All activities and resources interrelated.	Very good. Network shows interrelationship of all activities and resources.
2. Subcontracting	Poor. Does not show trade interrelationships.	Good. Interrelates with other resources.	Very good. Interrelates with all other resources.
3. Purchasing	Poor. Does not integrate with other resources.	Good. Interrelates with other resources.	Very good. Interrelates with all other resources.
4. Plant labor	Poor. Does not integrate with other resources.	Good. Interrelates with other resources.	Very good. Interrelates with all other resources.
5. Customer changes and selections	No good.	No good.	Very good. Integrates with all other resources.
Output Information			
1. Management summary information.	Good. Requires close examination.	Good, but complicated.	Very good. Provides daily summary analysis.
2. Program status and progress reports	Good.	Good, but fails to show incremental activity progress.	Very good. Shows incremental activity progress.
3. Timeliness and quality of danger signals.	Fair.	Fair. Semimonthly and monthly reports.	Excellent. Flags problem areas daily.
Usage to Date	Short period planning and continuous manufacturing planning. A fundamental control technique in manufacturing.	Primarily used for one-of-a-kind heavy construction projects. Very limited use in home building field.	Designed and developed for specific use in home building operations. Successfully used on all types and sizes of home building programs.

     c. Indicating the scheduled starts, workdays, and completion dates for subcontractor work on each job
     d. Showing the actual start dates, workdays, and completion dates for subcontractors on each job
     e. Indicating the number of days currently ahead or behind schedule for subcontractors on each job
     f. Showing the plan of work and schedule of performance for each subcontractor over the life of the project
     g. Permitting the subcontractor to plan and schedule his work load more effectively
     h. Helping the subcontractor better integrate his work with that of other subcontractors
     i. Building a more reliable subcontractor team
     j. Indicating the number of subcontractor trades required for the construction of each job
     k. Indicating the number of workdays required for each subcontractor trade to complete work on each job
     l. Indicating the number of crews needed by each subcontractor for concurrent construction operations
     m. Showing the location and availability of subcontractor trades for crash scheduling on troublesome jobs
     n. Indicating the lead time required for ordering subcontractors to report for work on each job
3. Purchasing—provides an efficient purchasing system for management by:
     a. Formulating a scheduling plan for the ordering and control of all procurement requirements
     b. Indicating the lead time required for ordering shipments of materials to job
     c. Integrating purchase order scheduling with construction operations scheduling
     d. Indicating schedule dates for placing purchase orders and receiving supplies at the construction site
     e. Permitting economical purchasing on a volume level rather than on a job ordering basis.
     f. Permitting a reduction in administrative effort and costs through bulk ordering
4. Inspections—presents management with an effective control technique to monitor inspection requirements by:
     a. Scheduling FHA or VA requirements for footing, foundations, cesspools, second and final inspections
     b. Scheduling town requirements for foundations, carpentry, and plumbing inspections

c. Scheduling company requirements for closing and final inspections
   d. Scheduling health department and water department for cesspool and water lateral inspections
   e. Scheduling lender compliance inspections for construction draws
   f. Scheduling customer final inspections for house acceptances
   g. Indicating lead time required for ordering inspections so construction will not be delayed
5. Plant labor—permits management to integrate and control the performance of plant labor on the project by:
   a. Indicating work activities to be performed by plant labor on each job
   b. Scheduling the work activities to be performed by plant labor
   c. Integrating plant labor activities with those of subcontractor trades on each job
   d. Planning the flow of plant labor from one job to the next on the project
6. Changes and selections—permits management to maintain close control over all changes and selections by:
   a. Indicating lead time required for incorporating changes and selections into each job's construction plan
   b. Scheduling the dates when each change and selection must be processed
   c. Indicating the types of changes and selections which are acceptable from customers
   d. Integrating all changes and selections into the purchasing, subcontracting, and construction plans for each job

A comparative analysis of the operations control techniques now used by home builders (Gantt charts, critical path method, and program control charts) is shown in Fig. 1.8. A factor comparison is presented on the installation, operation, planning, and scheduling applications, output information provided, and the results obtained in their usage to date by home builders. The comparison illustrates the strengths and weaknesses of each of these management control techniques and reveals the specific advantages of the program control chart for planning, scheduling, and controlling home building operations.

CHAPTER 2

Developing Operations Plans

OPERATIONS PLANNING

Operations planning is concerned with the balancing and control of all organization activities and resources associated with the accomplishment of the overall construction program. It is the most fundamental function in the home builder's scheme of operations for achieving the end goals of the organization.

Preliminary to the development of a home builder's operations plan, a thorough evaluation must be made of the functional capabilities of each department in the organization involved in its preparation. Their individual capabilities and responsibilites must be assessed in light of the essentials that each is to contribute to the operations planning function:

1. Marketing:
    a. Sales volume, product mix, and price range
    b. House styles, construction changes, and customer selections
    c. House models, furnishings, and site publicity
    d. Sales forecast and customer delivery requirements
    e. Sales program and promotional effort

Developing Operations Plans

2. Finance:
   a. Financial receipts and disbursements
   b. Financial resources and flow
   c. Financial balance and control
3. Construction:
   a. Construction capacity and resource requirements
   b. Construction and resource flow plans
   c. Construction work calendars and performance schedules
   d. Construction evaluation and control tools

The operations planning function represents a logical process through which the home builder may carefully consider and evaluate the capabilities of all departmental functions in the organization, and weigh their individual contributions toward achieving the overall goals of the construction program. It ensures that the functional capabilities and physical resources required by the organization to reach its goals are planned and projected on a total program basis. The marketing plan, which is the basis for operations planning, must be supported with realistic forecasts of market demand and a soundly conceived marketing program. The financial plan must reflect a realistic cash flow schedule to sustain the continuous flow of construction activities and resources required for the construction program. The construction plan must indicate how the anticipated construction work load will be balanced with overhead capacity and manpower.

Construction and resource plans must be developed to facilitate the management and control of anticipated increases in the construction work load. Target goals and schedule dates must be established for reaching each critical phase of the operations plan, from the start through the completion of the construction program. Management must then implement evaluation and control measures in order to monitor overall operations performance and ensure that intermediate targets are reached as scheduled so that the end goals of the program will be realized as planned.

Fundamentals of Planning

There are fundamental procedural actions which must be followed in the development of operations plans for a home building program. These actions involve:

1. Establishing planning goals to be reached in the construction program
2. Determining through requirements planning what construction activities will be involved and resources required for the construction program

3. Developing detailed construction activity and resource flow plans which serve as guidelines for reaching the goals of the construction program

Planning goals, the most basic requirement for operations planning, establish the limits or levels of the home builder's construction plan. They take into consideration the sales forecast, cash flow needs, and plant labor, subcontractor, material, and equipment requirements for the program. Goals are designed to coordinate and balance all contributions to the construction program, from each department within the organization and from outside sources. They are developed on the basis of the most efficient and effective utilization possible of all activities and resources required for the accomplishment of its sales goal. They establish definite objectives to be realized by the organization within a given period of time. In addition, they make possible the reduction of the operations cycle for the construction of housing units, with an increase in total unit output and a corresponding gain in profits. In turn, this reduces expenditure levels for interest, maintenance, theft, damage, storage, and other overhead items in the construction program.

Requirements planning, the next procedural requirement in operations planning, involves the home builder's determining in advance of actual construction what specific activities and resources will be required to carry out the construction program. It is concerned with the preplanning of the plant labor, subcontractor, and inspection activities that are to be performed in the construction of the housing units and deciding who will provide the services required for their performance. It is also concerned with determining what permits, materials, equipment, and other construction resources will be required to carry out the program and who will furnish them. It involves the advance planning of what structural changes (if any) will be permissible during construction, and what selection options will be available to customers.

Advance planning makes it possible for the home builder to determine in detail, prior to construction, what activities are to be performed and what resources will be needed for their performance. This enables the home builder to establish standard specification lists for all activities to be performed and all resources required on each housing unit. It also permits the identification of potential activity and resource problem areas in advance of their development, and provides some measurement of their probable effect on the construction program. In addition, it minimizes the possibility of delays arising from a lack of materials or the absence of subcontractors. Most important, it stimulates forward thinking by the home builder and leads to the development of realistic construction plans to accomplish program objectives.

Developing Operations Plans

Construction and resource flow plans translate the home builder's planning goals and requirements planning into an overall scheme of operations. They provide the guidelines required by the home builder for achieving the operations goals of the construction program. The construction flow plan is a logical means for diagramming, analyzing, integrating, measuring, and controlling the flow of all work activities involved in the construction program. It represents a logical scheme of operations for the construction of all housing units in the construction program in the least time possible. In addition, it provides a basis for scheduling, controlling, expediting, and monitoring the performance of all work activities in the construction program.

The resource flow plan, on the other hand, enables management to diagram, analyze, integrate, measure, and control the flow of all resources required in the construction program. It represents a schematic means for integrating the flow of all resource requirements into the mainstream of construction operations. This in turn provides the basis for scheduling, controlling, expediting, and monitoring the flow of all resources required in balance with the flow of all activities performed in the construction program.

Prerequisites for Planning

Prior to the actual development of the operations plans for the construction program, the home builder must have available basic operations information for their formulation. This information provides the basis upon which the goals, requirements, and detailed construction and resource plans of the program may be established by the home builder. The preliminary information prerequisite to planning includes:

1. Construction volume: the anticipated number of housing units in the construction program
2. Construction period: the estimated period of time required to carry out the construction program
3. Construction type: whether the housing units are to be of frame, concrete, brick, or of other construction
4. Construction design: ranch, colonial, split-level, etc.
5. Construction operations: custom, semicustom, etc.
6. Construction site: land improved or not, zoning requirements, availability of utilities, etc.
7. Construction features: basements, floor levels, garages, and any other special house feature
8. Construction models: whether house models are to be constructed, their number, type, etc.

9. Construction starts: whether construction starts are speculative, or on customer order
10. Construction changes: the type of construction changes allowed customers (if any)
11. Construction selections: the selection options available to customers
12. Construction financing: arrangements with lending institutions and construction draws required
13. Pricing: the price range for the finished houses

While this background information is not all-inclusive, nor designed to satisfy the financial and market planning needs of the home builder, it does provide the preliminary information required for launching the operations planning function of the construction program. The paragraphs which immediately follow illustrate the prerequisite information to be accumulated for developing a home builder's operations plan. The remaining sections of the book will present a practical, step-by-step demonstration of how, with this basic information, the systems management approach is used for planning, scheduling, and controlling a home builder's construction program.

Let us assume for the purpose of our demonstration that the home builder is a seasoned operator, constructing subdivisions of approximately fifty semicustom homes a year, on land that has already been improved, with streets laid out and utility lines installed. The construction product mix consists of three basic house styles: a two-story colonial, a split-level, and a rancher. The colonial is the most popular house style and has accounted for two-thirds of past sales. Models are available for customer inspection in each of the basic house styles, and there are several design variations in each model from which customers may choose their preferences. In addition, there are a number of structural changes and selection options which they may elect to incorporate in their homes.

The houses are of frame construction, and for the most part their exteriors are wood siding, with the exception being brick and stone veneer. Lot sizes average approximately one-third of an acre. The terrain is level and wooded, and the site is cluster-zoned. All utilities are available and ready for house hookup as required. Full basements are included, and oil-fired hot-water heaters with baseboard radiation systems are provided. Each style house has a slated foyer entranceway and a family room with a raised hearth fireplace. The number of bedrooms in each varies from three in the rancher to five in the colonial. The rancher has $1\frac{1}{2}$ baths and the colonial, $2\frac{1}{2}$. Formal dining rooms and large-sized kitchens set the mode for spacious living in all houses, and two-car garages are standard in each. Lawns are seeded, sodded,

Developing Operations Plans 33

and shrubbed to enhance the exterior of the homes and the appearance of the community.

House prices range from $28,000 for the ranch houses to $35,000 for the colonial-styled homes. Construction starts are made on customer orders. Conventional construction financing is used, and construction draws for work completed are to be made in accordance with the following schedule: 30 percent of the construction loan when the house is rough enclosed, 30 percent of the construction loan when fully enclosed, 20 percent of the construction loan when plumbing fixtures are installed, and 20 percent of the construction loan when the house is substantially completed. Additional charges for customer changes and extras are addended to the mortgage at the time of title closing with customers.

The construction volume, product mix, architectural styles, structural characteristics, and other pertinent operating details for the home builder have been established and defined. The remaining sections of the text will demonstrate the actual development of the operations plan, the establishment of operations schedules, and the implementing of operations controls for the home builder's construction program.

REQUIREMENTS PLANNING

Requirements planning is concerned with the definition and specification of all construction activities and resources directly associated with the development of the overall operations plan for carrying out the construction program. The activities and resources involved in the requirements planning function are:

1. Construction manpower
2. Construction materials
3. Changes and selections
4. Permits and inspections

Before entering the construction planning stage, the home builder must determine specifically what construction labor, subcontractors, material, equipment, permits, and inspections are required and what customer changes and selections will be permitted. The architect's plans and specifications for each model house must be analyzed to determine what activities are to be performed and what resources must be furnished for their construction. The home builder must decide what construction activities are to be carried out by the plant labor force and which activities are to be subcontracted; what materials and supplies are to be purchased by the company and which are to be furnished by subcontractors; what equipment is required and who will furnish it; what permits and inspections are required, who will furnish them, and when they are needed; and what changes and selections may be allowed customers without restricting the flow of construction operations.

The process of requirements planning enables the home builder to determine in advance of actual construction operations what specific activity and resource requirements are essential for implementing the construction program. In addition, it makes possible the standardization of construction activities and resource needs for each type of housing unit for methodizing their flow through the construction operations system.

Construction Manpower

The most important decision to be made by the home builder at this stage of planning is to determine what construction work is to be subcontracted and which activities are to be carried out by the plant labor crew. This decision would be based on the individual home builder's preferred business practice. Because of technical competence, pride in detailed craftsmanship, or the potential of greater earning capacity, one builder may prefer to perform his own masonry, framing, or finishing. On the other hand, because of proven managerial competence in handling large subcontractor teams, a lack of concern for technical details, or for financing considerations, another home builder may prefer to subcontract as much construction work as possible.

Let us assume that the home builder in this case has traditionally subcontracted all the major construction activities and has maintained a minimum work force for the performance of minor construction activities. Based on previous construction experience and the particular requirements of the houses that are to be built, the home builder prepares the following work list for the plant labor crew:

Plant Labor Work List

1. Clear and grub lots
2. Install window wells
3. Install slate walks
4. Install stone driveways
5. Install bath mirrors
6. Install medicine cabinets
7. Install shower doors
8. Clean site after first deck installed
9. Clean site after dry wall installed
10. Clean site after interior trim installed
11. Clean site and house after resilient floors installed

Developing Operations Plans 35

The plant labor work list represents a standard work specification for all construction activities to be performed by the plant labor crew on each house. It provides the means by which the home builder may direct, evaluate, and control the performance of plant labor.

Having decided which construction activities are to be performed by the plant labor crew, the next task of the home builder is to determine what subcontractor trades will be required for the construction program. Further study of the construction plans and specifications reveals that the following trades will make up the subcontractor team:

1. Surveyor
2. Excavator
3. Mason
4. Plumber
5. Rough carpenters
6. Roofer
7. Electrician
8. Electric utility company
9. Insulator
10. Drywaller
11. Painter
12. Ceramic tiler
13. Hardwood floor layer
14. Finish carpenters
15. Landscaper
16. Decorator
17. Resilient floor layer

Now that it has been decided which trades are required to make up the subcontractor team, the home builder must next determine what specific construction activities are to be performed by each team member. This information is determined through a work takeoff analysis of the construction plans and specifications, aided by the construction experience and judgment of the home builder. This leads to the preparation of a subcontractor work list, as follows:

Subcontractor Work List

1. Surveyor:
    Survey lot for stakeout
2. Excavator:
    a. Excavate foundation
    b. Backfill foundation
3. Mason:
    a. Install house foundation, cellar, and garage slabs
    b. Install chimney and fireplace
    c. Install patio foundation and slab
    d. Install foyer and hearth slate
4. Plumber:
    a. Install plumbing groundwork, water, and sewer lines
    b. Rough in plumbing and heating

    c. Install furnace, oil tank, water heater, heat grills, plumbing fixtures, trim, and plumbing appliances
5. Rough carpenter:
    a. Set steel joists, frame, and sheath house
    b. Set stairs; install windows, exterior doors, millwork, hardware, wood siding, and garage doors
6. Roofer:
    a. Install asphalt roof shingles
    b. Install gutters and spouts
7. Electrician:
    a. Rough in electricity
    b. Install plugs, switches, and light fixtures
    c. Hook up electrical appliances
8. Electric utility company:
    Connect electric service
9. Insulator:
    Insulate walls and ceilings
10. Drywaller:
    Hang dry wall, tape, and spackle
11. Painter:
    Paint exterior and interior
12. Ceramic tiler:
    Install ceramic bath tile
13. Hardwood floor layer:
    a. Install hardwood floors
    b. Finish hardwood floors
14. Finish carpenters:
    Install interior trim, doors, wall panels, shelving, kitchen cabinets, and vanities
15. Landscaper:
    Seed, sod, and shrub grounds
16. Decorator:
    Hang wallpaper
17. Resilient floor layer:
    Lay resilent floors

    The subcontractor work list is a standard work specification which outlines what construction activities are to be performed by which trades in the construction program. With the plant labor work list, it provides the basis for the subsequent integration of all work activities into a construction-flow plan to carry out and achieve the overall objectives of the construction program.

Developing Operations Plans

Construction Materials

At this stage of requirements planning, the home builder must determine what construction materials, supplies, and equipment are needed to build the houses and which of these items are to be subcontractor- or company-furnished.

Let us assume that the home builder's subcontractors have traditionally provided the following materials, supplies, and equipment with their services:

1. Surveyor: surveying equipment
2. Excavator: earth-moving equipment
3. Mason: blocks, brick, mortar, forms, and related supplies and equipment
4. Plumber: heating and plumbing equipment, fixtures, accessories, supplies, and related equipment
5. Roofer: roof shingles, gutters, spouts, and related supplies and equipment
6. Electrician: wiring, plugs, switches, and related accessories, supplies, and equipment
7. Insulator: insulation material and installation equipment
8. Drywaller: wallboard, cement, and related supplies and equipment
9. Painter: prime and finish paints and related supplies and equipment
10. Ceramic tiler: ceramic tile, cement, and related supplies and equipment
11. Hardwood floor layer: wood floors, finishes, and installation equipment
12. Landscaper: seed, sod, shrubs, and related supplies and equipment
13. Decorator: wallpaper and related supplies and equipment
14. Resilient floor layer: floor tiles, adhesives, and related equipment

A construction material takeoff analysis of the architect's plans and specifications reveals that the home builder must furnish the following items:

Construction Material Purchase List

1. Concrete
2. Basement windows
3. Lintels and lally columns
4. Window wells
5. Steel girders
6. Framing lumber
7. Sheathing lumber
8. Windows
9. Exterior doors and frames
10. Stairs
11. Exterior millwork
12. Exterior hardware
13. Wood siding
14. Garage doors

15. Interior doors and frames
16. Interior trim and shelving
17. Wall paneling
18. Slate and crushed stone
19. Vanities
20. Kitchen appliances
21. Kitchen cabinets
22. Finish hardware
23. Bath mirrors
24. Medicine cabinets
25. Shower doors
26. Plumbing appliances

The construction materials requirements list provides the home builder with a standard material specification guide for the procurement of all company-furnished items required for the construction of each house. Once these requirements have been established, it is possible to integrate their procurement flow into the construction activity flow for the program.

Changes and Selections

Customer changes and selections are potential problem areas in the construction program unless they are standardized so that they may be methodically incorporated into the flow of construction operations. The more customized the house, the greater the home builder's need for preplanning what changes and selections will be allowed to customers.

Let us assume that our semicustom home builder has elected to make the following changes and selections available as customer options:

Changes and Selections List

1. Masonry:
    a. Fireplace design
    b. Slate colors
2. Carpentry:
    a. Window styles
    b. Door styles
    c. Millwork styles
    d. Siding types
    e. Wall panel types
    f. Interior trim style
3. Bathrooms:
    a. Mirror styles
    b. Medicine cabinet styles
    c. Shower door types
    d. Plumbing fixture colors
    e. Ceramic tile colors
    f. Vantity types

Developing Operations Plans

4. Appliances:
    a. Kitchen appliance types and colors
    b. Plumbing appliance types and colors
5. Floors:
    a. Hardwood floor types and finishes
    b. Resilient floor types and styles
6. Painting:
    a. Interior colors
    b. Exterior colors
7. Decorating:
    Paper patterns
8. Roofing:
    Shingle colors
9. Kitchen cabinets:
    Cabinet styles and finishes
10. Light fixtures:
    Fixture types and styles

The construction changes and selections list represents a standard specification guide for customer options which provides the home builder with a logical basis for planning, scheduling, and controlling their processing into the flow of construction operations.

Permits and Inspections

Another requirement which must be planned in advance of actual construction so that the flow of operations will proceed smoothly is the building permits and inspections that will be needed for the houses In addition to any special permits and inspections required by the individual subcontractors, the home builder must provide the following:

Permits and Inspections List

1. Building permits:
    a. Construction permit: obtained prior to construction
    b. Plumbing permit: obtained prior to construction
    c. Fireplace permit: obtained prior to construction
    d. Occupancy permit: obtained prior to occupancy
2. Town inspections:
    a. Foundation inspection: when the foundation is installed
    b. First carpentry inspection: when the house is framed and sheathed
    c. Second carpentry inspection: when the interior trim is completed
    d. First plumbing inspection: when the plumbing is roughed in
    e. Second plumbing inspection: when the plumbing is completely installed

f. Electrical inspection: when the electrical work is completely installed
   g. Final house inspection: when the house is completed and ready for occupancy
3. Lender inspections:
   a. First inspection: when the house is rough enclosed
   b. Second inspection: when the house is completely enclosed
   c. Third inspection: when the kitchen cabinets are installed
   d. Final house inspection: when the house is completed and ready for occupancy
4. Customer inspections:
   Final house inspection: when the house is completed and ready for customer acceptance and occupancy

The permits and inspections requirements list, like the previously mentioned construction requirements list, provides the home builder with a standard specification guide for planning and integrating the flow of this activity into the construction operations plan.

CONSTRUCTION PLANNING

Construction planning involves the logical integration of all construction activities and resources required for carrying out the home builder's construction program. It is concerned with both the preparation and the interrelationships of each of the following basic elements of the construction program:

1. Construction-time estimates
2. Construction-flow plan
3. Resource lead time
4. Resource-flow plan

The construction plan must be designed to accommodate the most complex and frequently constructed housing unit in the home builder's product line. It must be schematically formulated so that it will provide the most effective processing in the least time possible for the flow of all housing units through their construction cycle. In the case of our semicustom home builder, the construction plan would be based on the construction requirements of the two-story colonial house, the largest and most frequently constructed housing unit in the product line.

Construction-time Estimates

The first step to be taken by the home builder in the development of the construction plan is to establish time estimates for each work

Developing Operations Plans

activity to be performed in the construction of the colonial unit. Time estimates must be prepared for all plant labor, subcontractor, and inspection activities, each of which consumes direct construction time and effort in the construction of the house. Collectively, they account for the total direct work involved in its construction, and the effectiveness of their integration determines the length of time required for its completion.

PLANT LABOR
ACTIVITY TIME ESTIMATES

Description of Activity	Estimated Workdays
Plant Labor	
1. Clear and grub lots	2.0
2. Install window wells	1.0
3. Install slate walks	1.5
4. Install stone driveway	2.0
5. Install mirrors	.5
6. Install medicine cabinets	.5
7. Install shower doors	1.0
8. Clean site after first deck installed	1.0
9. Clean site after dry wall installed	1.0
10. Clean site after trim installed	1.0
11. Clean house and site after resilient floors installed	4.0

Figure 2.1

Developing the overall construction plan on the basis of the time estimates required for work activities involved in building the largest house in the product line will automatically create slack time in the overall construction program. Though the time differentials between the performance of work activities on the ranch and colonial houses will not be of major significance, it will ease the flow of labor and subcontractor work activities between the houses under construction. This slack is desirable as it provides a built-in time factor against the unforeseen contingencies which are inherent in home building operations.

The time estimates for plant labor activities shown in Fig. 2.1 are based on the home builder's experience with the plant labor crew and his estimated time for the performance of each of their work activities. The subcontractor activity time estimates in Fig. 2.2 are based on esti-

SUBCONTRACTOR ACTIVITY TIME ESTIMATES

Description of Activity	Estimated Workdays
Surveyor	
Survey for stakeout	2.0
Excavator	
1. Excavate foundation	2.0
2. Backfill foundation	1.0
Mason	
1. Dig and pour foundation footings	2.0
2. Install foundation walls	4.0
3. Parge foundation walls	1.0
4. Tar foundation walls	1.0
5. Install cellar slab	2.0
6. Grade for garage slab	1.0
7. Install garage slab	1.0
8. Install garage apron	1.0
9. Install chimney fireplace	3.0
10. Dig and pour patio footings	1.0
11. Install patio foundation	1.0
12. Install patio slab	1.0
13. Lay foyer and hearth slate	2.0
Plumber	
1. Install plumbing groundwork	2.0
2. Install water and sewer lines	2.0
3. Rough in plumbing	3.0
4. Rough in heating	2.0
5. Set furnace, oil tank, and water heater	2.0
6. Install heat grills and registers	2.0
7. Install plumbing fixtures	3.0
8. Install plumbing appliances and trim	1.5
Rough Carpenters	
1. Set steel joists	1.0
2. Install first deck	2.5
3. Frame first floor and garage	3.5
4. Install second deck	2.0
5. Frame second floor	3.0
6. Frame roof	3.0
7. Sheath house	3.0
8. Block out	2.0
9. Set stairs	2.0
10. Set exterior door frames and windows	1.5

Figure 2.2

SUBCONTRACTOR
ACTIVITY TIME ESTIMATES (*Continued*)

Description of Activity	*Estimated Workdays*
11. Install exterior doors.	1.5
12. Install exterior millwork.	3.0
13. Install wood siding.	3.0
14. Install garage doors.	1.0
Roofer	
1. Install asphalt roof shingles.	3.0
2. Install gutters and spouts.	1.5
Electrician	
1. Rough in electricity.	2.0
2. Install plugs and switches.	1.5
3. Hook up kitchen appliances.	2.0
4. Install light fixtures.	2.0
Electric Utility Company	
Connect electric service to house.	1.0
Insulator	
Insulate walls and ceilings.	2.0
Drywaller	
Hang dry wall, tape, and spackle.	5.0
Painter	
1. Paint house exterior.	4.0
2. Paint house interior.	3.0
Ceramic Tiler	
Install ceramic bath tile.	2.0
Hardwood Floor Layer	
1. Lay hardwood floors.	4.0
2. Finish hardwood floors.	3.5
Finish Carpenters	
1. Install interior doors, panels, shelving, and trim.	4.0
2. Install kitchen cabinets.	2.5
3. Install vanities.	2.0
Landscaper	
Seed, sod, and shrub grounds.	3.0
Decorator	
Hang wallpaper.	3.0
Resilient Floor Layer	
Install resilient floor tile.	2.0

mates furnished by the subcontractors, with adjustments made by the home builder to reflect actual experience with their work performance. The time estimates in Fig. 2.3 for job inspections are based on the home builder's experience with the lender and town inspectors and past customers.

JOB INSPECTION
ACTIVITY TIME ESTIMATES

Description of Activity	Estimated Workdays
Town Inspections	
1. Foundation inspection—foundation installed	1.0
2. First carpentry inspection—house framed and sheathed	1.5
3. Second carpentry inspection—interior trim completed	1.0
4. First plumbing inspection—plumbing roughed in	1.0
5. Second plumbing inspection—plumbing installation completed	1.0
6. Electrical inspection—electrical installation completed	1.0
7. Final house inspection—house completed, ready for occupancy	1.0
Lender Inspections	
1. Rough enclosure completion inspection	1.0
2. Full enclosure completion inspection	1.0
3. Kitchen cabinet installation inspection	1.0
4. Final house inspection—house completed, ready for occupancy	1.0
Customer Inspection	
Final house inspection—for acceptance	1.0

Figure 2.3

Construction-flow Plan

The home builder has determined at this point what plant labor, subcontractor, and inspection activities must be performed for the construction of the largest unit in his product line. In addition, realistic time estimates have been established for the accomplishment of each of these construction activities.

The next step to be taken by the home builder is to tie together all these construction activities into a construction-flow plan. This will enable the builder to identify the interdependencies and interrelationships of all work activities involved in the construction process and to establish the logical sequence in which they are to be performed. The construction-flow plan is developed with the use of the critical path network technique, which provides the builder with a logical means

Developing Operations Plans

for integrating all work activities to be performed into a construction-flow plan for his building program.

Preparation of the construction-flow plan is simplified with the use of a program control work chart, as shown in Exhibit II. The work chart presents the home builder with a practical means for rough drafting the construction-flow plan and integrating into this plan the flow of all orders placed for changes, selections, subcontractors, purchases, and inspections. It is construction workday oriented, having a time scale which spans the top of the chart and measures the number of workdays required for construction, from stakeout to completion.

The grid in the center of the chart facilitates the preparation of the construction-flow network and also its evaluation after completion. Only actual construction workdays are considered when preparing the network, and no allowances are made for weekends, holidays, vacations, and other days on which construction work is not planned.

The program control work chart shows the construction-flow plan for the home builder's largest construction job, the colonial house. Construction activities, the sequence in which they are to be performed, start and completion events, slack time between construction activities, the critical activity path, and the concurrent activity work paths, are each represented by the following network symbols:

○   The circle represents either the start or the completion of a work activity and is called an "event."

③   With a number in it, the circle indicates the numerical sequence of the starting or completing event, reading from left to right on the network.

→   The solid arrow represents a work activity to be performed, and the arrowhead indicates its activity flow. A brief description of the work activity involved is posted above the activity arrow.

⇢   The broken-line arrow represents slack time which may exist between work activity arrows. It indicates that no work is required between these work activities. Slack arrows appear only on concurrent activity paths.

The *critical path*, which travels from left to right through the center of the network, is the longest uninterrupted series of work activity arrows in the construction-flow plan. It determines the overall length of time required for the construction of the house, from the starting event on the first day to the completion event on the final workday.

Concurrent paths are the shorter series of work activity arrows which travel parallel to the network's critical path. The concurrent work paths are terminated with slack arrows which tie them back into the mainstream of construction activities, the critical path.

Inasmuch as all the required information for its preparation has already been accumulated, the actual plotting of the construction-flow plan is simply a mechanical process for the home builder. The following three items provide the basic data needed:

1. Plant labor activities and their time estimates
2. Subcontractor activities and their time estimates
3. Inspection activities and their time estimates

What remains to be done is the actual plotting of each of the construction activities on the central grid section of the work chart. With each of the activity listings and their time estimates at hand, the home builder commences to plot on the grid each of the activities proportionate in length to their time estimates as measured against the chart's construction workday scale. Circles are drawn to indicate the beginning event before stakeout and the subsequent events which follow as the construction work activities are plotted in the form of arrows. Event numbers are not entered in the circles until the construction-flow network has been completed and their numerical sequence firmly established.

As the activity arrows are drawn on the chart, their work descriptions are posted directly above them, and they are checked off the activity lists from which they were obtained. Thus, by the same logical process employed by the home builder for the actual construction of the house, from stakeout to foundation excavation, to digging and pouring footings, installing the foundation, and on through structural completion, the construction-flow plan is developed, arrow by arrow, event by event, until the completion event is finally reached.

The work chart provides the home builder with the means for preparing a rough draft of his construction-flow plan and for sounding out its logic before it is put into actual use. The rough draft of the network will meander across the chart as the builder thinks out the logic of the sequential flow of each construction activity and plots it. As the construction-flow plan develops, the critical (the longest) path in the network will become evident, and the concurrent paths with their slack will be apparent. Just where the slack arrows will appear or how much slack time they will have cannot be determined until all the work activities have been plotted on the chart. As the slack time between the concurrent and critical paths does become apparent, the concurrent paths are tied back into the critical path with slack arrows. After it has been completed, the critical path network on the work chart is copied in finished form on the program control chart, where its structural form,

Developing Operations Plans

activity flow, and numerical sequencing of events for the construction-flow plan are finalized.

Once the construction-flow plan is completed, the home builder may prepare from its format a construction activity plan as shown in Fig. 2.4. The construction activity plan:

1. Lists all construction activities to be performed in their sequence, from stakeout to customer acceptance
2. Indicates the number of workdays required for the performance of each construction activity
3. Indicates those activities which are on the critical path and the number of workdays required for their performance
4. Shows the cumulative number of construction workdays that accrue for critical path activities from stakeout and clear lot to completion and customer occupancy

The construction activity plan provides the home builder with a yardstick against which actual construction performance in the field may be measured against the construction-flow plan.

Resource Lead Time

The term "resource lead time" refers to the number of construction workdays that must be allowed between the time that construction changes, selections, materials, subcontractors, and inspections are ordered and the time that they are needed at the site so that construction operations will not be delayed.

If the construction-flow plan is to function smoothly and provide the efficiency in construction operations for which it was designed, it is of major importance that the home builder determine at this stage the lead time required for ordering all resources. Inasmuch as customer changes and selections have a multiple effect on the ordering of other resources, as well as upon the construction-flow plan, their lead times must receive prime attention. This requires an evaluation of the changes and selections that are available to customers. What specific options are offered? When are customers notified that these options are available? When are these options exercised by customers? When are these changes and selections incorporated into the construction-flow plan? How much lead time is allowed for construction labor and subcontractors to implement these changes and selections once they are received from customers?

The home builder has already determined during the requirements planning stage what changes and selections are to be available to customers. The construction-flow plan indicates the event and construction

CONSTRUCTION ACTIVITY PLAN
SEMICUSTOM HOME CONSTRUCTION

Work Activity Number	Description of Activity	Total Workdays	Critical Path Workdays	
			No.	Cum.
1-2	Stakeout, clear lot	2.0	2.0	2.0
2-3	Excavate foundation	2.0	2.0	4.0
3-4	Dig and pour footings	2.0	2.0	6.0
4-5	Install foundation walls	4.0	4.0	10.0
5-6	Foundation inspection	1.0		
5-7	Parge foundation walls	1.0		
5-9	Install plumbing groundwork	2.0	2.0	12.0
5-10	Install water and sewer lines	2.0		
7-8	Tar foundation walls	1.0		
8-11	Backfill foundation	1.0		
9-12	Grade garage floor	1.0		
9-14	Pour cellar slab	2.0	2.0	14.0
11-13	Install window wells	1.0		
12-15	Pour garage slab	1.0		
14-16	Set steel girders	1.0	1.0	15.0
15-17	Pour garage apron	1.0		
16-18	Install first deck	2.5	2.5	17.5
18-19	Clean up debris	1.0		
18-20	Frame first floor and garage	3.5	3.5	21.0
20-21	Install second deck	2.0	2.0	23.0
21-22	Frame second floor	3.0	3.0	26.0
22-23	Install chimney and fireplace	3.0		
22-24	Frame roof	3.0	3.0	29.0
24-25	Sheath house	3.0	3.0	32.0
25-26	Carpentry inspection	1.5		
25-27	Set exterior door, window frames	1.5		
25-28	Set furnace, tank, water heater	2.0		
25-29	Set stairs	2.0		
25-30	Block out	2.0		
25-31	Rough in heating	2.0	2.0	34.0
25-32	Shingle roof	3.0		
25-33	Rough in plumbing	3.0		
27-34	Install exterior doors	1.5		
31-35	Rough in electricity	2.0	2.0	36.0
34-39	Install exterior millwork	3.0		
35-36	Connect electric service	1.0		
35-37	Plumbing inspection	1.0		
35-38	Install insulation	2.0	2.0	38.0
38-42	Install dry wall	5.0	5.0	43.0

Figure 2.4

CONSTRUCTION ACTIVITY PLAN
SEMICUSTOM HOME CONSTRUCTION (Continued)

Work Activity Number	Description of Activity	Total Workdays	Critical Path Workdays	
			No.	Cum.
39-40	Apply wood siding	3.0		
40-41	Install garage doors	1.0		
41-43	Dig and pour patio footings	1.0		
41-50	Paint house exterior	4.0		
42-44	Clean up debris	1.0		
42-46	Install plugs and switches	1.5		
42-47	Install heat registers and grills	2.0		
42-48	Install ceramic bath tile	2.0		
42-51	Lay hardwood floors	4.0	4.0	47.0
43-45	Install patio foundation	1.0		
45-49	Pour patio slab	1.0		
48-53	Install plumbing fixtures	3.0		
50-52	Install gutters and spouts	1.5		
51-55	Install interior doors and trim	4.0	4.0	51.0
52-54	Landscape grounds	3.0		
54-58	Install slate walks	1.5		
55-56	Carpentry inspection	1.0		
55-57	Clean up debris	1.0		
55-59	Paint house interior	3.0	3.0	54.0
58-60	Install stone driveway	2.0		
59-61	Install vanities	2.0		
59-62	Install kitchen appliances	2.0		
59-63	Install foyer and hearth slate	2.0		
59-64	Install kitchen cabinets	2.5	2.5	56.5
59-65	Interior decorate	3.0		
64-66	Install mirrors, medicine cabinets	1.0		
64-67	Install shower doors	1.0		
64-68	Install plumbing appliances and trim	1.5		
65-69	Install light fixtures	2.0	2.0	58.5
69-70	Electric inspection	1.0		
69-71	Plumbing inspection	1.0		
69-72	Finish hardwood floors	3.5	3.5	62.0
72-73	Install resilient floors	2.0	2.0	64.0
73-74	Final town inspection	1.0		
73-75	Final bank inspection	1.0		
73-76	Final customer inspection	1.0		
73-77	Customer occupancy permit	1.0		
73-78	Clean house and grounds	4.0	4.0	68.0
78-79	Customer acceptance and occupancy	2.0	2.0	70.0

workday numbers when the changes and selections must be incorporated by plant labor and subcontractors into construction operations. With this information, the home builder can proceed to prepare the changes and selections lead-time schedule shown in Fig. 2.5, by posting the

CHANGES AND SELECTIONS
LEAD-TIME SCHEDULE

Event Number	Description	Workday Required	
		For Construction	From Customer
22	Fireplace—design	27	17
25	Windows—type	33	17
25	Roof shingles—color	33	17
27	Doors—type	34	17
34	Millwork—type	36	17
39	Siding—type	39	17
41	Exterior paint—colors	43	32
42	Ceramic tile—colors	44	32
42	Hardwood floors—type	44	32
48	Plumbing fixtures—colors	46	32
51	Wall panels—type	48	32
51	Interior trim—type	48	32
54	Exterior slate—colors	51	32
55	Interior paint—colors	52	32
59	Foyer, hearth slate—color	55	32
59	Vanities—type	55	32
59	Kitchen appliances—type	55	32
59	Kitchen cabinets—type	55	32
59	Interior decoration—paper	55	32
64	Mirrors—type	57	47
64	Medicine cabinets—type	57	47
64	Shower doors—type	57	47
64	Plumbing appliances—type	57	47
64	Light fixtures—type	57	47
69	Hardwood floors—finish	59	47
72	Resilient floor tile—pattern	63	47

Figure 2.5

event numbers, descriptions, and workday numbers on which they are required, so that construction may proceed without delay.

Customers are notified what changes and selections they may make when the sales agreement is signed and before actual construction is

Developing Operations Plans

started. They are furnished customer option forms which list the changes and selections offered and indicate when they must be made. As illustrated in Fig. 2.5, the home builder establishes three points on the construction work calendar on which changes and selections must be finalized by customers: the seventeenth, thirty-second, and forty-seventh construction workdays. This allows sufficient lead time for customers to submit their changes and selections to the builder for inclusion in material and subcontractor orders.

After the changes and selections lead-time schedule has been established, the home builder must next determine the lead time required for placing purchase orders to ensure that materials are available at the construction site when needed. During the requirements planning stage, the company-furnished materials required for construction were determined. An examination of the construction-flow plan reveals at what event and construction workday numbers materials must be at the site for plant labor and subcontractor use to sustain the flow of construction operations. With this information, the home builder can prepare the purchase order lead-time schedule shown in Fig. 2.6. The event numbers are posted on the schedule form in the sequence in which the materials are required, followed by a description of the materials and postings indicating the construction workdays that they are needed at the site.

The next requirement in the preparation of the purchase order lead-time schedule is to determine how much lead time is required for the actual delivery of each material to be ordered. After this is decided, the home builder must select several points on the construction workday calendar where purchase orders may be economically released in lots so that a number of material requirements may be ordered at the same time. The construction workdays on which the purchase orders for materials are to be released are then posted in the last column on the schedule form. The purchase order lead-time schedule ensures that materials will be ordered early enough to permit the continuity of construction activities in accordance with the construction-flow plan.

Following the preparation of the purchase order lead-time schedule, the home builder must establish the lead-time schedule for ordering subcontractors to report for the performance of their trades. The construction activities to be carried out by the subcontractors were determined during the requirements planning stage by the home builder. The construction-flow plan indicates at what event and construction workday number each subcontractor must report at the site for work. This information provides the home builder with sufficient data to initiate preparation of the subcontractor lead-time schedule shown in Fig. 2.7. Event numbers are posted on the schedule in the sequence in which the subcontractors are to first report for work at the site, followed by a brief notation of the work to be performed and an entry indicating the construction workday on which they must report. The number of lead-time days required for the subcontractors to report for work is then determined, after which several ordering points are designated on the construction workday calendar. The home builder then posts on the schedule the construction workday that each subcontractor is

PURCHASE ORDER
LEAD-TIME SCHEDULE

Event Number	Description	Workday Number	
		Required at Site	Day Ordered
0	Building permits	1	Before stakeout
3	Concrete	5	Before stakeout
4	Basement windows	7	Before stakeout
4	Lintels	7	Before stakeout
11	Window wells	14	9
14	Steel girders	15	9
14	Lally columns	15	9
16	Lumber	16	9
25	Windows, doorframes	33	18
25	Stairs	33	18
27	Exterior doors	34	18
34	Exterior millwork	36	18
34	Exterior hardware	36	18
39	Wood siding	39	18
40	Garage doors	42	33
51	Interior trim, doors, shelving	48	33
51	Wall panels	48	33
54	Exterior slate	51	33
58	Crushed stone	53	33
59	Vanities	55	33
59	Kitchen appliances	55	33
59	Kitchen cabinets	55	33
59	Finish hardware	55	33
64	Mirrors	57	48
64	Medicine cabinets	57	48
64	Shower doors	57	48
64	Plumbing appliances	57	48
73	Customer occupancy permit	65	65

Figure 2.6

Developing Operations Plans

SUBCONTRACTOR
LEAD-TIME SCHEDULE

Event Number	Description	Workday Number	
		Required at Site	Day Ordered
1	Surveyor—survey lot	1	Before stakeout
2	Excavator—dig foundation	3	Before stakeout
3	Mason—install foundation	5	Before stakeout
5	Plumber—groundwork, laterals	11	Before stakeout
8	Excavator—backfill	13	9
9	Mason—cellar, garage slabs	13	9
14	Rough carpenters—frame and sheath	15	9
22	Mason—chimney, fireplace	27	18
25	Plumber—rough plumbing and heating	33	18
25	Roofer—shingle roof	33	18
31	Electrician—rough electricity	35	18
35	Electric company—connect service	37	33
35	Insulator—insulate	37	33
38	Drywaller—board, tape, and spackle	39	33
41	Mason—install patio	43	33
41	Painter—exterior and interior paint	43	33
42	Electrician—plugs, switches	44	33
42	Plumber—heat grills, registers	44	33
42	Ceramic tiler—bath tile	44	33
42	Hardwood floorer—lay floors	44	33
48	Plumber—fixtures	46	33
50	Roofer—spouts, gutters	47	33
51	Finish carpenters—interior trim	48	33
52	Landscaper—seed, sod, shrub	48	33
59	Finish carpenters—kitchen cabinets, vanities	55	48
59	Electrician—kitchen appliances, light fixtures	55	48
59	Mason—foyer, hearth slate	55	48
59	Decorator—interior decorate	55	48
64	Plumber—appliances, trim	57	48
69	Hardwood floorer—finish floors	59	48
72	Resilient floorer—lay floors	63	48

Figure 2.7

to report for work on each major construction activity. Sufficient lead time is allowed in the ordering of the subcontractors to ensure that they have adequate time to plan for their work at the site, so that they may be able to report as scheduled and the flow of their construction activities will proceed as planned.

The job inspection lead-time schedule illustrated in Fig. 2.8 is prepared in a manner similar to that followed for each of the other resource lead-time schedules. This schedule establishes the lead time required

JOB INSPECTION
LEAD-TIME SCHEDULE

Event Number	Description	Workday Number	
		Required at Site	Workday Ordered
5	Town foundation inspection	11	8
25	Town carpentry inspection	33	30
25	Lender rough enclosure inspection	33	30
35	Town plumbing inspection	37	34
42	Lender full enclosure inspection	44	41
55	Town final carpentry inspection	52	49
64	Lender kitchen cabinet inspection	57	54
69	Town final plumbing inspection	60	57
69	Town electrical inspection	60	57
73	Town final house inspection	65	62
73	Lender final house inspection	65	62
73	Customer house inspection	65	62

Figure 2.8

for the ordering of all lender, town, customer, and other job inspections which must be performed during the course of construction on each house. The job inspection activities needed were determined during the requirements planning stage for the eventual preparation of the construction-flow plan. Indicated on the flow plan are the event and construction workday numbers on which job inspections must be made. This information is posted on the job inspection lead-time schedule, entering in the event column the event numbers in the sequence in which the inspections are to be made, describing each inspection activity, and recording the construction workdays on which they are required. After determining the number of lead-time days required for each job inspection, the construction workdays on which they are to be ordered are then posted on the schedule form. The job inspection lead-time

Developing Operations Plans 55

schedule enables the home builder to plan the integrated flow of all job inspections into the construction-flow plan.

Another resource lead-time schedule that should be prepared by the home builder is that for the flow of construction draws which follow the rough enclosure, full enclosure, kitchen cabinet, and final job inspections made by the lending institution. Shown in Fig. 2.9 is the home builder's construction draw lead-time schedule, which coincides with the job inspections made by the lender prior to its release of construction funds. While the inspections required by the bank are already covered

CONSTRUCTION DRAW
LEAD-TIME SCHEDULE

Event Number	Description	Workday Number	
		Day Requested	Day Ordered
25	Rough enclosure draw	33	30
42	Full enclosure draw	44	41
64	Kitchen cabinet draw	57	54
73	Final draw	65	62

Figure 2.9

in the job inspection lead-time schedule, construction draws are of sufficient importance to the builder to warrant the preparation of a separate lead-time schedule for their control.

Inasmuch as it is assumed that plant labor is readily available for deployment when and where needed at the construction site, there is no need for the preparation of a lead-time schedule for its ordering and allocation. However, the home builder does need a work plan that indicates at what event in the construction-flow plan plant labor will be needed, for what purpose, and on what construction workday. The work activities to be performed by plant labor were determined in the requirements planning stage and subsequently incorporated in the construction-flow plan. With this information, the builder can prepare the plant labor work activity plan shown in Fig. 2.10, which illustrates where and when plant labor activities are to be integrated into the construction-flow plan.

Resource-flow Plan

Operations flow planning is subdivided into two major areas: construction-flow planning, which has already been discussed and demonstrated, and resource-flow planning, which will be described and illustrated in the material that follows.

Construction-flow plans provide management with the schematic means required for coordinating and controlling the flow of all direct construction activities involved in the construction of housing units. Resource-flow plans, on the other hand, schematically integrate the flow of all resources into the mainstream of construction operations. The home builder's prime management function is to oversee the efficient flow of both construction activities and resources and to interlock their flows into an integrated system of operations.

PLANT LABOR
WORK ACTIVITY PLAN

Event Number	*Description of Activity*	*Workday Ordered*
1	Clear lot, stakeout	1
11	Install window wells	14
18	Clean up debris after first deck installed	18
42	Clean up debris after dry wall installed	44
54	Lay slate walks	51
55	Clean up debris after interior trimmed	52
58	Install stone driveway	53
64	Install mirrors	57
64	Install medicine cabinets	57
64	Install shower doors	57
73	Final clean up, house and grounds	65

Figure 2.10

Resource-flow planning starts at the requirements planning stage when the home builder determines what labor, material, subcontractors, inspections, and other resources are required to carry out the construction program. The planning of resource flow is further developed when the resource lead-time schedules are prepared indicating when resource items are required for the construction-flow plan and when they must be ordered so as to not delay the flow of construction operations. The resource-flow plan is fully developed when the flow of resources has been interwoven into the construction-flow plan by the home builder.

The network planning technique used for designing and developing the construction-flow plan provides the schematic means necessary for intermeshing both the construction activity and resource plans. Through the use of the program control work chart illustrated in Exhibit II and

Developing Operations Plans

previously discussed, the actual integration of the construction-flow plan and the resource-flow plan is simplified.

Shown in the lower half of the work chart is the home builder's resource-flow plan which integrates changes and selections, purchase orders, subcontractor orders, labor orders, and job inspections into the construction-flow plan. The information required for plotting the resource-flow plans appears on the home builder's resource lead-time schedules. Resource items are plotted on the chart in their individual resource categories. Vertical arrows indicate what resource items are to be ordered on what construction workdays. When the resource items are actually required is indicated by the construction activity arrows which appear on the construction-flow plan. Thus, through this simple yet very effective process, the flows of both construction operations and resource requirements have been interlocked into a totally integrated operations system for the home builder.

CHAPTER 3

Establishing Operations Schedules

OPERATIONS SCHEDULING

The operations scheduling function may be defined as the process of regulating the orderly integration and flow of all construction activities and resources through the operations cycle by means of the systematic ordering and release of all manpower and material required under a routine plan which utilizes the total organization most effectively.

Operations scheduling involves the synchronization of construction activity and resource requirement schedules and their subsequent integration into a master timetable for scheduling the flow of all construction manpower and material through the operations process. It provides the means for keeping track of all activities and resources integral to the flow of construction operations, charting the progress of their ordering and subsequent utilization in the operations process. It furnishes a method by which schedules and delivery dates may be altered to optimize the flow of overall construction operations. While operations planning establishes the requirements for performance in the organization, operations scheduling provides the measurement function to maintain overall performance within the limits of these requirements.

Custom Flow

The complexity of the scheduling function varies considerably with the nature and scope of the home builder's construction program. Scheduling is most complex in the custom-order home building operation (intermittent orders) and comparatively simple in the continuous-job home building operation (flow orders). Many home building operations are actually a combination custom- and continuous-job operation (semi-custom). Scheduling systems must be designed to reflect the individual needs of home builder firms.

In the custom-job operation, the focus is on the individual customer order, its delivery date, the construction operations which must be performed, the allocation of resources and time to it, and the control of progress against its schedule. This is *custom-order flow scheduling*.

On the other hand, when we have a continuous-job construction operation characterized by a large number of standardized (or semistandardized) units, close scheduling control of individual job orders is not necessary. This is termed *production-flow scheduling*.

In a custom-order scheduling system, individual plans and specifications are made up for each job order. Each order is scheduled through construction operations separately. Work loads imposed on plant labor and subcontractors are considered for each separate job. Work load schedules are prepared for bottleneck operations, and assumptions are made that these critical areas of activity set the pace for all other activities. Through continuous expediting, the progress of the order is tracked to meet individual customer delivery commitments.

Production Flow

When close scheduling control is not required for home building operations as it is in the case of individual custom-order processing, then production-flow processing methods are used. The close control over individual order processing is not required because each job order is similar to preceding ones. In production-flow processing, all orders flow through a fixed sequence of operations. This sequence of operations is based on the most complex job produced in the product line, with limited structural variations. Instead of scheduling the separate flow of each individual job order, the entire construction operation is on a scheduled production-flow basis. All job orders processed are automatically coordinated with the overall operations schedule through the predetermined design of the operations-flow system.

Construction scheduling is practically complete when the level of construction operations is established in the overall operations plan. Construction control is achieved by merely adjusting the construction levels in accordance with sales experience. This adjustment of construction level is an adjustment of the rate of flow, and is termed *production-flow control*. The rates of flow of construction manpower and material at the construction site are adjusted to match the required flow of finished housing units. Plant labor, subcontractors, and suppliers must then maintain these flow rates.

Supply contracts must contain volume flexibility within certain limits so that the operations site can adjust supply rates to maintain control over material requirements and inventories. Repeat orders for materials may be replaced with blanket supply contracts for furnishing material needs over the extended period of the construction program. The control over supplies must be precise because the delay of materials can hold up the entire sequence and flow of operations on the construction program.

CONSTRUCTION SCHEDULING

Construction scheduling is concerned with the orderly regulation of the flow of housing units through the overall operations process. It requires the methodical timing of the flow of all construction activities and resources required for the production of housing units in accordance with their preconceived plan of operations flow. It provides a timetable for the measurement and monitoring of the rates of flow of all manpower and material requirements into the operations process and the flow rate of housing units through the operations system.

The construction scheduling process involves the establishment of time limits for the flow of each job order through operations. The job flow rate for semicustom home building operations is based on the flow time required for the construction of the most complex unit in the product mix. The structural design differentials between the least and the most complicated units to be constructed are controlled within practical limits of variation so that a smooth rate of construction flow may be maintained. The most complex structural unit is first scheduled through the operations system, which has been designed to accommodate this particular job unit. The scheduling of this complex unit automatically establishes the maximum time rate for the flow of all subsequent job orders through the operations system. The differential in construction complexity between individual units in the product mix provides operations slack time which eases the schedule flow of all units through operations in accordance with the overall timetable for the construction program.

Establishing Operations Schedules

Job Schedules

Through the use of the program control work chart (see Exhibit III), the preparation of the construction schedule for the first job unit to be produced is a simple matter. In addition to providing the construction-flow plan by which the job order is to be processed and the resource-flow plan for materials and other requirements, the work chart makes provision for the actual scheduling of the job through its entire cycle of construction operations.

The length of time required for the construction of the most complex semicustom job has been established in the home builder's construction-flow plan (seventy construction workdays). Directly beneath the construction workday scale on the work chart, provision is made for posting the construction work dates on which the job is to be processed through operations. The first figure shown on the construction workday scale is zero and is included on the scale to signal those resource items which must be ordered or available for use at the construction site before the actual start of job operations. Starting with this figure, calendar work dates are posted in the horizontal schedule squares in chronological sequence, from the designated prestart date through the final date required for construction. To illustrate, if the prestart date selected is March 1, it is posted in the schedule square beneath the zero on the workday scale. Disregarding those calendar dates on which no construction work is planned (weekends, holidays, vacations, etc.), the work dates for construction are entered from a calendar in chronological sequence in the horizontal job schedule column for the job on the work chart.

The work chart enables the home builder to completely schedule all construction activities and resources required for construction of the job unit in several minutes. Posting the time schedule for the flow of the job through the operations system automatically provides schedule dates for seventy-eight direct construction activities that must be performed and for over one hundred resource items that must be ordered for the construction of the job. Actually, there are over 450 bits of schedule information (scheduled starts, performance, and completion dates) associated with the construction of the semicustom home shown on the work chart. Once the scheduled construction work dates have been posted, all schedule information for the job becomes immediately available for instant visual readout from the chart by the home builder. Schedule dates are automatically provided for the flow of all job activities through the operations system and the flow of all job resource requirements into the mainstream of construction operations.

Production Schedules

There are three fundamental approaches to production scheduling that are applied in home building operations. They are:

1. Sequential-flow scheduling
2. Parallel-flow scheduling
3. Concurrent-flow scheduling

The difference between these approaches to production scheduling is that in *sequential-flow scheduling* one job cannot be started without completing the construction of the preceding job; in *parallel-flow scheduling* a number of jobs are started, constructed, and completed at the same time; and in *concurrent-flow scheduling*, a combination of sequential- and parallel-flow scheduling, a number of jobs are under construction at the same time but in different stages of construction.

To more clearly distinguish the differences between these scheduling approaches, their applications will be illustrated. Let us assume that the home builder decides to construct the fifty semicustom homes only one at a time. That is, one home will be gradually completed, then another home started and completed, and so on until all homes in the construction program are completed. The home builder's goal is to complete construction of one home at a time, thus extending the overall operations cycle for the program over a long period of time. This type of scheduling would be unnecessary and uneconomical.

Let us assume that, instead, the home builder decides to construct all fifty semicustom homes simultaneously, scheduling their production flow in parallel. The same construction activity would then be performed on each home at the same time before the following construction activity would begin. Under this method of scheduling, each of the fifty homes would be in the same stage of construction at any given time inasmuch as they were all started at the same time and will be completed at the same time. While it is possible for the home builder to construct the homes in production lots, it would be impractical to attempt their total production in one lot because of the limited labor, materials, finance, and management resources available for such a large undertaking.

The third approach considered by the home builder is concurrent-flow scheduling. With this scheduling method, a number of semicustom homes would be under construction at the same time, but each would be in a different stage of construction. As the construction program progressed, each home would be gradually constructed, from start to completion. At each instant of time, different construction activities would be performed on each house under construction. Concurrent-flow scheduling has proved to be the most effective method of scheduling production in home building operations. This is the approach selected by the home builder in our demonstration.

Having decided that the fifty semicustom homes are to be produced concurrently, the home builder must next determine the feasible number of construction starts that may be scheduled each week. Based on previous sales experience, the present local housing demand, and projected sales demand, the builder tentatively plans two construction starts a week. With adequate labor at hand, subcontractors available, and no foreseeable material shortages, the builder firms up his production plan to start construction on two jobs a week. At this level of production, it is concluded that if the first job is ready to start on March 1, the construction program will be completed by December 1.

The home builder is prepared at this point to establish schedule work dates for the production of each of the fifty jobs in the construction program. The planning and scheduling information that has been plotted on the program control work chart (Exhibit III) is transferred to the program control chart shown in Exhibit IV. This chart is designed to accommodate the concurrent scheduling of all fifty jobs in the home builder's construction program. After plotting the work chart material on the program control chart, the job numbers are posted in the sequence in which they are to be started in the Job Number column at the left of the chart. This establishes the priority of the flow of the construction starts in the program. As customer orders are received, their section and block numbers are addended to the sequential job number already posted in the column.

The actual posting of the construction schedule dates for each job is simplified with the use of the program control calendar shown in Fig. 3.1. The calendar shows the number of construction workdays available during each month of the construction program. Weekends, holidays, vacations, and other anticipated nonworkdays are omitted. The calendar year and months are posted in the left-hand columns, the construction work dates are posted in the grid section, and the holidays to be observed are noted in the right-hand column.

In accordance with the planned schedule for two construction starts a week, the home builder next posts in the vertical schedule column adjacent to the Job Number column the prestart calendar dates for each job at the rate of two units a week, which are, chronologically, 3/3, 3/8, 3/10, 3/15, 3/17, etc. Referring to the program control calendar, the home builder then plots in the horizontal schedule columns for each job the work dates for their construction in chronological sequence, until they have been completely scheduled. When all of the calendar work dates have been posted for the construction of each job, the concur-

		Program Control Calendar Construction Workdays		
Yr.	Mo.	1 2 3 4 5 6 7 8 9 10 11 12 13 14 15 16 17 18 19 20 21 22 23		Holidays
66	Mar.	3/1 3/2 3/3 3/4 3/5 3/8 3/9 3/10 3/11 3/12 3/15 3/16 3/17 3/18 3/19 3/22 3/23 3/24 3/25 3/26 3/29 3/30 3/31		
66	Apr.	4/1 4/2 4/5 4/6 4/7 4/8 4/9 4/12 4/13 4/14 4/15 4/16 4/19 4/20 4/21 4/22 4/23 4/26 4/27 4/28 4/29 4/30 /		
66	May	5/3 5/4 5/5 5/6 5/7 5/10 5/11 5/12 5/13 5/14 5/17 5/18 5/19 5/20 5/21 5/24 5/25 5/26 5/27 5/28 / / /		
66	June	6/1 6/2 6/3 6/4 6/7 6/8 6/9 6/10 6/11 6/14 6/15 6/16 6/17 6/18 6/21 6/22 6/23 6/24 6/25 6/28 6/29 6/30 /		
66	July	7/1 7/2 7/6 7/7 7/8 7/9 7/12 7/13 7/14 7/15 7/16 7/19 7/20 7/21 7/22 7/23 7/26 7/27 7/28 7/29 7/30 / /		7/4 Independence Day
66	Aug.	8/2 8/3 8/4 8/5 8/6 8/9 8/10 8/11 8/12 8/13 8/16 8/17 8/18 8/19 8/20 8/23 8/24 8/25 8/26 8/27 8/30 8/31 /		
66	Sept.	9/1 9/2 9/3 9/7 9/8 9/9 9/10 9/13 9/14 9/15 9/16 9/17 9/20 9/21 9/22 9/23 9/24 9/27 9/28 9/29 9/30 / /		9/6 Labor Day
66	Oct.	10/1 10/4 10/5 10/6 10/7 10/8 10/11 10/12 10/13 10/14 10/15 10/18 10/19 10/20 10/21 10/22 10/25 10/26 10/27 10/28 10/29 / /		
66	Nov.	11/1 11/2 11/3 11/4 11/5 11/8 11/9 11/10 11/12 11/15 11/16 11/17 11/18 11/19 11/22 11/23 11/24 11/26 11/29 11/30 / / /		11/11 Veteran's Day 11/25 Thanksgiving

Figure 3.1

rent-flow schedules for all fifty semicustom homes will have been established.

RESOURCE SCHEDULING

While construction scheduling is concerned with the regulation of the flow of construction units through their preconceived plan of operations, resource scheduling involves the establishment of timetables for the ordering of all labor, materials, inspections, and other requirements for maintaining the continuous flow of construction operations. The flow of all resources must be scheduled to coincide with the flow of all construction activities involved in the actual construction of each job. The timetables must be as carefully synchronized as the flow networks upon which their activity and resource plans are based. Their thorough integration is essential to the realization of the overall construction operations objectives.

Subcontractor Trades

Once the construction schedules have been posted for each job on the program control chart, the lead-time ordering and activity performance schedules for subcontractors are automatically established.

Establishing Operations Schedules

Posted in the resource plan at the bottom of the chart is the subcontractor lead-time schedule which shows what subcontractor groups are to be ordered and indicates with vertical arrows the construction workday and date they must be ordered for each job. For instance, the surveyor, excavator, mason, and plumber must be ordered by the home builder for Job 1 before stakeout, on or before March 1. There must be sufficient lead time allowed for the subcontractors to plan their work loads and to be able to report on the dates scheduled for their performance. The dates on which they must be available at the job site to perform their work is indicated in the vertical schedule date column that appears directly above and below their work activities plotted on the construction-flow plan.

When the subcontractors are ordered by the home builder, they are advised of the jobs on which they are to report for work and of the dates on which they are to perform their activities as established by the subcontractor work schedule shown in Fig. 3.2. In addition to indicating the project name, its location, and other pertinent information, the schedule shows the jobs and the scheduled start and completion dates for the performance of the subcontractor's work activities on each. The work schedule also provides for entries to be made by the subcontractor relating to his actual performance against schedule on each job: whether work was started on each as scheduled or the number of days performed ahead or behind schedule. The Comments column provides for special instructions from the home builder to the subcontractor concerning work activities on each job or for notations by subcontractors regarding actual or potential work problems that may be confronted on each.

Shown to the right of the construction-flow plan on the chart is a construction activity starts directory, which indicates the event numbers in the plan at which each construction activity to be performed on each job is to start. The directory provides a reference table for the home builder's guidance, showing what subcontracting, labor, and inspection activities are to start at what specific event numbers in the construction-flow plan.

Plant Labor

Inasmuch as the home builder maintains direct control over the plant labor force, there is no need for lead-time scheduling their work activities as required for the subcontractor trades.

The work activities to be performed by the plant labor crew are shown in the plant labor schedule of the resource plan on the program control chart. The workdays and dates on which they are to be performed are indicated with vertical arrows. The vertical schedule date columns which appear directly above and below the work activity arrows on

SUBCONTRACTOR WORK SCHEDULE

Project_____

Location_____

Date_____

Subcontractor_____
Address_____
Work Description_____

Job No.	Schedule Dates		Variance +/− Days		Comments
	Start	Finish	Start	Finish	

Work Schedule

Figure 3.2

Establishing Operations Schedules

the construction-flow plan indicate when the work is to start and be completed on each job in the construction program.

Purchase Orders

The purchase order lead-time schedule shown in the resource plan section of the chart indicates what materials must be ordered on what workdays and dates so that they may be available when needed at the construction site.

The items to be ordered are grouped in purchase lots to limit the frequency of placing purchase orders. The home builder may further consolidate his purchasing by ordering a number of the purchase items in each group for the needs of several or more jobs. The vertical arrows indicate the lead-time workdays and dates by which purchase orders must be placed for each item so as not to delay the flow of construction operations. The schedule dates on which they are to be ordered are indicated in the vertical schedule columns immediately above and below the purchase order arrows in their resource plan.

Changes and Selections

The changes and selections lead-time schedule appears in the resource plan section of the program control chart. The schedule indicates what change and selection orders must be obtained from customers and the lead-time dates by which they must be received by the home builder for construction operations to proceed as planned.

To facilitate the follow-up action needed for the receipt of the changes and selections, they are subdivided into four schedule groups. A description of the changes and selections required is posted in each group, and the construction workday by which they must be received at the latest is indicated with vertical arrows. The vertical schedule columns immediately above the arrows indicate the latest schedule dates by which all changes and selections must be received from customers.

Job Inspections

In the resource plan section of the chart is an inspection lead-time schedule which indicates what inspections are required on each job and when they must be ordered by the home builder.

The job inspections posted in the lead-time schedule are those required by local building agencies, by the lending institution, and for customer acceptance of jobs when construction has been completed. Vertical arrows posted alongside each inspection requirement indicate the construction workdays on which they must be ordered. The vertical schedule column immediately above the inspection arrow indicates the schedule dates for ordering the inspections.

CHAPTER 4

Implementing Operations Controls

PRODUCTION CONTROL TECHNIQUES

The production control function involves continuous coordination and corrective action by management over the entire operations system to ensure the smooth flow of construction activities and resource requirements both into and through the system.

Production coordination calls for continuous follow-up action on the release and flow of all purchase orders, work orders, customer orders, customer complaints, customer selections, and structural changes. The ordering and actual performance of all labor, subcontractor, and inspection work activities requires constant appraisal and review, and their performance schedules must be continuously evaluated and improved where possible. The effective flow of all construction manpower and materials into the mainstream of operations and the efficient flow of all construction jobs through the operations system requires continuous expediting over the entire construction program.

Follow-up action ensures management that the flow of construction activities and resource requirements will coincide with the schedule performance needs of overall production operations. Expediting the flow of operations not only reduces the job construction cycle, but also facilitates the growth of sales by accelerating the availability of homes over a given period. This, in turn, creates an atmosphere of action which is stimulating to both salesmen and customers.

There are a number of production control tools available to management for evaluating and controlling the flow of production operations: sales forecasts, construction-flow plans, requirements plans, resource-flow plans, construction schedules, resource schedules, and lead-time schedules. Thus, management has at its disposal the essential tools required for maintaining firm control over production operations and for achieving the overall objectives of the construction program. Production control makes it possible for management to systematically coordinate and direct all operations to ensure that jobs are completed as scheduled, their standards of quality are acceptable, and they are constructed at reasonable cost.

The best method for evaluating the effectiveness of operations is schedule performance. It provides management with a practical yardstick for gauging how well the organization is coordinated, the satisfaction of customers, and the smoothness of the overall flow of operations. This yardstick of management may be used for measuring:

1. Job schedule performance
2. Subcontractor schedule performance
3. Supplier schedule performance
4. Labor schedule performance
5. Changes and selections schedule performance
6. Job inspection schedule performance

While management's prime concern is overall operations performance, its effectiveness to a major extent depends upon the individual performance of plant labor, subcontractors, suppliers, inspection agencies, and the processing of customer changes and selections. Unless the schedule performance of each of these essential elements of the operations system meets overall operations requirements, management cannot be assured that program objectives will be realized. Only through continuous follow-up action may management be reasonably confident that production is under complete control and that operations objectives will be reached as planned.

Construction Status Evaluation

To ensure that overall operations flow smoothly in accordance with established plans and schedules for the construction program, the home builder must systematically:

1. Review all jobs under construction to determine their current work status
2. Measure the amount of construction progress that has been realized on each job
3. Record the work activities completed and the construction status of each job on the program control chart
4. Evaluate the work performed and determine the work activities yet to be performed for the completion of each job
5. Evaluate the resources made available and determine those yet required for completing the construction of each job

Production control in home building involves the methodical review, measurement, recording, evaluation, and coordination of overall construction operations to ensure that all necessary construction activities are performed and essential resource requirements are available as scheduled.

While it is important that each job under construction be inspected daily to monitor and supervise the quantity and quality of construction work in progress, it is equally important for management review and control purposes that the overall construction status of each unit be evaluated at least once a week. This involves determining what specific stages of construction each job is in, what work has been performed and materials furnished on each, and what construction work and materials are presently required to complete construction. Construction status evaluation is accomplished through management's weekly audit of all jobs under construction, recording the present work status of each job and plotting and measuring construction performance on the program control chart.

The job status record shown in Fig. 4.1 facilitates documenting at the site the specific stage of construction of each job. Listed on the form is a description of the job work activities to be performed as shown in the construction-flow plan on the program control chart. Provision is made in columns adjacent to the work activity listing for posting the job numbers of units that are in particular work activity stages. The job status record makes it possible for management to record during inspection the specific work activity status and the exact construction stage reached by each job.

After the construction status for each job has been posted on the job status record, this information must be plotted on the program control chart, as shown in Exhibit V. The construction progress realized on each job is plotted with colored markers on the transparent acetate sheet which overlays the chart. Two progress bars are drawn over the horizontal schedule columns for each job. The top progress bar refers

Implementing Operations Controls

to the critical work activities completed on the job as measured by the critical path on the construction-flow plan shown in the center of the chart. The lower progress bar relates to the most delinquent concurrent work activity finished on the job as measured on the concurrent work paths of the construction-flow plan.

There is only one critical path of operations on the construction-flow plan against which the progress of critical work activities on each job may be measured. However, at times there may be a number of different concurrent paths on which work is progressing in parallel with work on the critical path. When a number of separate concurrent path work activities are being performed at once, credit for their construction progress plotted on the chart applies to the most delinquent of the concurrent path work activities. The purpose of this is to avoid overlooking concurrent path work activities not yet performed, which may otherwise be credited as completed.

After the construction status has been posted and the construction progress bars plotted for each job, the program control chart will provide the home builder with a visual size-up of the overall status of all jobs in the construction program. In terms of production flow the chart will show:

1. The specific stage of construction that each job has reached
2. The length of construction time required to complete each job
3. The relative progress of construction on each job
4. Those jobs yet to be started and constructed
5. The critical and concurrent work activities that have been completed on each job
6. The critical and concurrent work activities yet to be performed on each job

In addition, the chart will provide significant information on the flow of resources into construction operations in terms of:

1. Subcontractor trades now available at the construction site and the jobs they are working on
2. Subcontractor trades that must be ordered and available to complete each job
3. Construction materials that must be ordered and available for each job
4. Plant labor that must be available for work on each job
5. Job inspections which must be ordered for jobs not completed
6. Changes and selections which must be processed on jobs not yet completed
7. The flow of labor and subcontractors between construction jobs

JOB STATUS RECORD					
Work Activity Number	Work Activity Description	Number of Workdays	Work Status	
			Job	Job
1-2	Stakeout, clear lot	2.0		
2-3	Excavate foundation	2.0		
3-4	Dig and pour footings	2.0		
4-5	Install foundation walls	4.0		
5-6	Foundation inspection	1.0		
5-7	Parge foundation walls	1.0		
5-9	Install plumbing groundwork	2.0		
5-10	Install water and sewer lines	2.0		
7-8	Tar foundation walls	1.0		
8-11	Backfill foundation	1.0		
9-12	Grade garage floor	1.0		
9-14	Pour cellar slab	2.0		
11-13	Install window wells	1.0		
12-15	Pour garage slab	1.0		
14-16	Set steel girders	1.0		
15-17	Pour garage apron	1.0		
16-18	Install first deck	2.5		
18-19	Clean up debris	1.0		
18-20	Frame first floor and garage	3.5		
20-21	Install second deck	2.0		
21-22	Frame second floor	3.0		
22-23	Install chimney and fireplace	3.0		
22-24	Frame roof	3.0		
24-25	Sheath house	3.0		
25-26	Carpentry inspection	1.5		
25-27	Set exterior door and window frames	1.5		
25-28	Set furnace, tank, water heater	2.0		
25-29	Set stairs	2.0		
25-30	Block out	2.0		
25-31	Rough in heating	2.0		
25-32	Shingle roof	3.0		
25-33	Rough in plumbing	3.0		
27-34	Install exterior doors	1.5		
31-35	Rough in electricity	2.0		
34-39	Install exterior millwork	3.0		
35-36	Connect electric service	1.0		
35-37	Plumbing inspection	1.0		
35-38	Install insulation	2.0		
38-42	Install dry wall	5.0		

Figure 4.1

JOB STATUS RECORD (*Continued*)

Work Activity Number	Work Activity Description	Number of Workdays	Work Status	
			Job	Job
39-40	Apply wood siding	3.0		
40-41	Install garage doors	1.0		
41-43	Dig and pour patio footings	1.0		
41-50	Paint house exterior	4.0		
42-44	Clean up debris	1.0		
42-46	Install plugs and switches	1.5		
42-47	Install heat registers and grills	2.0		
42-48	Install ceramic bath tile	2.0		
42-51	Lay hardwood floors	4.0		
43-45	Install patio foundation	1.0		
45-49	Pour patio slab	1.0		
48-53	Install plumbing fixtures	3.0		
50-52	Install gutters and spouts	1.5		
51-55	Install interior doors and trim	4.0		
52-54	Landscape grounds	3.0		
54-58	Install slate walks	1.5		
55-56	Carpentry inspection	1.0		
55-57	Clean up debris	1.0		
55-59	Paint house interior	3.0		
58-60	Install stone driveway	2.0		
59-61	Install vanities	2.0		
59-62	Install kitchen appliances	2.0		
59-63	Install foyer and hearth slate	2.0		
59-64	Install kitchen cabinets	2.5		
59-65	Interior decorate	3.0		
64-66	Install mirrors, medicine cabinets	1.0		
64-67	Install shower doors	1.0		
64-68	Install plumbing appliances and trim	1.5		
64-69	Install light fixtures	2.0		
69-70	Electric inspection	1.0		
69-71	Plumbing inspection	1.0		
69-72	Finish hardwood floors	3.5		
72-73	Install resilient floors	2.0		
73-74	Final town inspection	1.0		
73-75	Final bank inspection	1.0		
73-76	Final customer inspection	1.0		
73-77	Customer occupancy permit	1.0		
73-78	Clean house and grounds	4.0		
78-79	Customer acceptance and occupancy	2.0		

8. Those subcontractors that are delayed due to material shortages or the delinquent performance of other trades

Construction Variance Control

The most important function performed by the home builder is the control function which calls for the systematic analysis and monitoring of the overall flow of construction operations. This requires constant evaluation of the flow of all construction units through the operations system and the flow into this system of all resources required for their production. Construction activity and resource flows must be continuously appraised by management in terms of their actual operations flow as opposed to their performance schedules.

Management's most effective method for measuring how well the overall construction program is functioning is by comparing the actual construction performance on each job with the construction schedules. The variance (difference) between actual and scheduled construction performance provides the home builder with a yardstick to measure the overall efficiency of production flow through operations and the effectiveness of the plans and schedules used in the construction program.

In order to maintain construction variance control over the scheduled flow of construction jobs through the operations system, the home builder must methodically:

1. Monitor the actual progress of each job by following up on the timely performance of all labor, subcontractor, and inspection activities and by expediting the ordering and delivery of all materials required for construction
2. Measure the number of construction workdays variance between actual and scheduled construction performance on each job
3. Determine the reasons for the construction variance between actual and scheduled performance, pinpointing those construction activities and resources which are causing actual performance to differ from that scheduled
4. Establish records to document construction variance between actual and scheduled performance on each job and to identify significant variance trends in the construction program
5. Maintain control over construction variance between actual and scheduled performance for each job through the continuous evaluation and expediting of those construction activities and resources which cause the variation from schedule

The program control chart shown in Exhibit V has a job status section for recording the variance that may develop between the actual and scheduled construction performance on each job. The job status section

Implementing Operations Controls 75

of the chart is reproduced in Fig. 4.2 to more clearly illustrate its format and contents. Entries are posted weekly on the clear surface of the chart with colored pencils, and erasures are made with a soft cloth or tissue. A description of the information furnished in the job status section of the chart is as follows:

1. At the top of the section is posted the week-ending date on which construction status entries are made.
2. In the first column is a list of the jobs on which construction status is being posted (in actual practice the job numbers are listed on the far left of the program control chart, not adjacent to the job status section).
3. The second column lists the critical path work activities now being performed or about to be performed on each job. The top horizontal construction progress bar shown on the chart for each job ends at a point that is parallel to the particular critical path work activity shown on the construction-flow plan in the center of the chart.
4. The third column lists the concurrent path work activities which lag behind other concurrent path activities and on which work is now being performed or is about to be performed on each job. The bottom horizontal construction progress bar shown on the chart for each job ends at a point parallel to the particular concurrent path work activity shown on the construction-flow plan.
5. Shown in the fourth column is the number of construction workdays variance between actual and scheduled performance on critical path work activities. Construction variance is measured in terms of workdays ahead (+) or behind (−) schedule. To illustrate, on 6/7/66 the top construction progress bar for Job 1 should have extended no further than the date 6/7 shown on the grid, to indicate that critical construction activities were performed as scheduled. Instead, the progress bar extended over 6/8, indicating that critical construction work had been performed a day ahead of schedule. Accordingly, the construction variance shown in column 4 (which relates to column 2) is +1, as actual work is one day ahead of schedule.
6. The fifth column shows the number of construction workdays variance ahead or behind schedule for each job on their respective concurrent work activity paths. On Job 25, for instance, the foundation backfill activity should not have been reached until 6/11, yet it was reached on 6/7, three construction workdays ahead of schedule. Accordingly, a construction variance of +3 is posted.

With the job status information shown on the chart, the home builder can measure the number of construction workdays variance between actual and scheduled performance on each job and determine which

JOB STATUS—DATE: 6/7/66

VARIANCE +/− WORKDAYS

Job No.	Critical Activities	Concurrent Activities	Critical Work	Concurrent Work
1	Finished	Finished	+1	+1
2	Cleanup	Final inspection	0	−3
3	Resilient floors	Final inspection	−1	0
4	Resilient floors	Final inspection	0	0
5	Finish hardwood floors	Electrical inspection	0	0
6	Light fixtures	Interior decorate	0	−1
7	Kitchen cabinets	Driveway	0	−1
8	Interior trim	Driveway	−2	+1
9	Interior trim	Plumbing fixtures	−1	−2
10	Install hardwood floors	Plumbing fixtures	−2	0
11	Dry wall	Exterior paint	−2	0
12	Dry wall	Ceramic tile	−1	+1
13	Dry wall	Wood siding	0	+1
14	Insulate	Exterior millwork	−1	0
15	Rough electricity	Rough plumbing	0	0
16	Rough heating	Set stairs	0	0
17	Sheath house	Shingle roof	0	0
18	Frame roof	Install chimney	0	0
19	Frame roof	Install chimney	+2	+2
20	Frame roof	Install chimney	+4	+4
21	Frame second floor	Install chimney	+4	0
22	Install second deck	Install chimney	+4	0
23	Install first deck	Clean up	+3	0
24	Set steel	Garage apron	+2	+2
25	Plumbing groundwork	Backfill	+2	+3
26	Foundation walls		+2	
27	Foundation walls		+2	
28	Dig footings		+2	
29	Excavate foundation		+3	
30	Stakeout		+4	

Figure 4.2

activities and resources are accelerating or delaying their completion. Further, it is possible to distinguish those activities that are critical and require special management attention for job construction to continue and would justify double-up work activities so as not to endanger overall construction schedules. The weekly revision of the job status data provides management with a continuing size-up of how effective construction performance is for all jobs scheduled.

The job status information posted on the program control chart is designed to provide management with a running account of construction performance and of necessity is revised and updated weekly to reflect current operating conditions in the field. It is important that the weekly construction variance information compiled for critical path work activities on each job under construction be documented on a permanent record for transmittal from the field to the main office to apprise top management of the current status of field operations. The form used for transmitting this information is the construction status report shown in Fig. 4.3. The report makes provision for recording weekly what jobs are under construction, their scheduled start and finish dates, their plus or minus performance variance from established schedules on the reporting date, and pertinent comments relating to the reasons for their variance from schedule.

With the information provided on the construction status report each week, the home builder may establish a construction variance control system with the construction schedule variance control form shown in Fig. 4.4. This form permits management to record the weekly performance variance from schedule for all jobs currently under construction. With this information management may discern the over- or under-schedule variance trends for each job in the construction program. In turn, management may weigh the realism of the plans and schedules established for the program against actual experience and the construction activities and resources yet required for achieving its objectives. Troublesome jobs may be isolated from others for special remedial action to improve their construction performance. Where it is evident that work trades and suppliers are not performing and delivering up to expectations, decisions may be reached and acted upon to resolve their particular problems.

PROGRAM CONTROL TECHNIQUES

The prime functions of production control are generating and expediting the uninterrupted flow of labor, material, and other resources into operations and the continuous flow of all construction jobs through

	CONSTRUCTION STATUS REPORT			
Project			Date	
Job No.	Schedule		Variance +/− Days	Comments
	Start	Finish		

Figure 4.3

CONSTRUCTION SCHEDULE VARIANCE CONTROL

Job No.	Schedule Finish	Weekly Performance Variance from Schedule								
		/	/	/	/	/	/	/	/	/

Figure 4.4

the operations system. The main purpose of program control, on the other hand, is to balance the efficiency of the resource flow into the operations system against the effectiveness of operations outputs in terms of finished construction jobs.

Operations efficiency depends in large measure upon the effectiveness of overall operations control as measured by sound planning, realistic schedules, limited shortages, accurate lead times, and reliable time intervals for construction. Planning provides the basis for operations control and allows management to weigh labor, material, and other needs against operations requirements for carrying out the construction program. The most essential element of operations planning is sales planning which, depending upon its degree of reliability, will minimize the fluctuations in operating levels and permit the optimum utilization of all organization capabilities. Scheduling should be based on realistic sales plans and known construction intervals, yet it should allow reasonable slack for unforeseen contingencies in both which may delay overall operations.

The most common deterrents to operations performance are labor and material shortages. The levels at which they may be tolerated vary widely among construction jobs and programs depending upon their individual complexities and the adequacies of their lead-time schedules. The need for crash operations may reflect unrealistic planning and scheduling; however, this may be unavoidable at times due to unforeseen subcontractor, supplier, or weather problems. At the other extreme, there undoubtedly will arise idle time in operations, a certain amount of which should be anticipated and considered normal. However, wide fluctuations between the extremes of crash operations and excess idle time must be guarded against by management. There must be a constant effort on the part of management to reach an optimum balance between the efficiency of all operating elements and the effectiveness of operations outputs toward realizing the end objectives of the construction program. A well-balanced construction program not only contributes greatly toward stabilizing overall operations, but in addition leads to increased productivity, more dependable resource flows, and improved customer relations.

Program Status Evaluation

In order to develop and maintain effective balance and control over the construction program, the home builder must continuously review and evaluate the overall flow of operations in the program. This demands constant appraisal of the balance between the flow of resources into operations and the flow of finished jobs through and out of operations. It further requires critical evaluations of unbalanced operation flows to determine their causes and to develop realistic solutions for their prompt correction.

Implementing Operations Controls

In addition to its applications for planning, scheduling, and controlling the flow of resources and production into and through operations, the program control chart provides a practical visual means for evaluating the effectiveness of management's control over the total construction program. Individually, the performance bars plotted for each job under construction measure the cumulative construction progress made against their construction-flow plans and performance schedules. Collectively, the performance bars for construction jobs present a total size-up of overall construction performance on the program. The terminal points reached by the performance bars for each job, when viewed in relation to each other, assume the configuration of a diagonal line extending from the lower left to the upper right of the chart. This diagonal line is referred to as the *program flow curve*, as it reflects the overall movement and progress of all jobs under construction in the program. The shape of the program flow curve provides an excellent measure of management's success in balancing and controlling the overall progress flow of the construction program.

The program flow curve may assume a number of different forms depending, again, on how well resource and job flows are balanced and controlled by management. The curve may take on the appearance of a straight line, convex line, concave line, S line, or an inverted S line, each of which is illustrated in Figs. 4.5 to 4.9. Or it may assume the appearance of a combination of these various curves. An analysis of the most common forms of program flow curves is presented in the material which immediately follows.

The straight-line program flow curve shown in Fig. 4.5 illustrates the configuration that the curve would assume on the program control chart for an extremely well-balanced and controlled construction program. The smooth production flow shown for all jobs under construction is the result of an effectively coordinated flow of all resources required for their construction. The high level of management control reflected

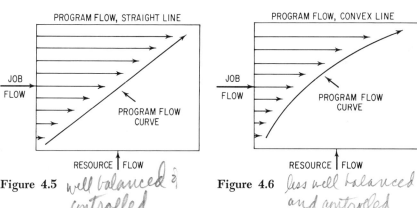

Figure 4.5 *well balanced & controlled*

Figure 4.6 *less well balanced and controlled*

in this program flow curve is a measure of the efficiency of the operations planning, scheduling, and control functions established for this particular construction program.

The convex-shaped program flow curve illustrated in Fig. 4.6 reflects a less well-balanced and controlled construction program than that shown in Fig. 4.5. The convex curve indicates that either management follow-up on operations is not properly applied across the production line or resources have been improperly planned and allocated. Both the start-up and the finishing trades are outpacing the performance of the structural trades that are engaged in the mid-operations phase. Possibly the structural trades are undermanned, lack materials for their performance, are hampered by bad weather, or have been improperly scheduled. The problem may rest with poor scheduling and control of the start-up and finishing trades. Whatever the reasons for the imbalance

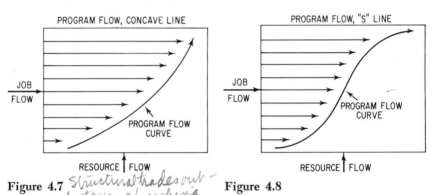

Figure 4.7   Figure 4.8

of operations flow, the program flow curve signals that immediate management action is required to stabilize the flow of overall operations.

The concave program flow curve shown in Fig. 4.7 indicates that the progress of the structural trades is outdistancing the performance of the start-up and finishing trades. The problem in this situation may be that management is applying too much follow-up action on the structural trade activities and not enough on the others. Or the problem may lie in poor planning and scheduling of work trade activities or their material requirements. Possibly the structural trades have with experience increased their work proficiency level above that of the other trades. In this case, rescheduling may be required to bring the production line back into balance. Management may be focusing too much attention on the quality and quantity of the work performed by the structural trades and not enough on the work of the start-up and finishing trades. There may be terrain problems which restrict the work activities of the start-up trades on the more recently started jobs. Here again,

Implementing Operations Controls

whatever the causes of the unbalanced flows of overall operations, the program flow curve calls management's attention to the problems that need immediate corrective action.

Another form of program flow curve which signals for management action is the S-shaped curve shown in Fig. 4.8. This form of curve may reveal that either construction starts are not fast enough or the finishing trades are not pacing the structural trades. There may be finishing problems which are restricting the smooth flow of jobs out of the structural stage of construction. Possibly, architectural requirements are hampering the performance of the finishing trades and may have to be modified to facilitate the job flow of the trades affected. Material shortages and absentee trades may be the problems restricting the effective flow of jobs through operations. Management may be contributing to the unbalanced job flow by pressing the performance of the finishing

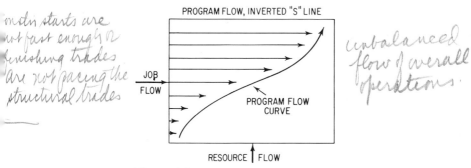

Figure 4.9

trades at the near-completion stage in order to step up title closings. Whatever the reasons for the S-shaped program flow curve, it is evident that management does not have effective control over the total flow of operations and must promptly apply remedial action.

Shown in Fig. 4.9 is an inverted S-line curve which also calls management's attention to an unbalanced flow of overall operations. This curve indicates that jobs are not moving rapidly enough from their start-up stage into the structural stage of operations, possibly because of poor planning and scheduling, a shortage of critical materials, subcontractors not being available, or some other special reason. On the other hand, the structural trades may have increased their productivity, or their work may have been expedited more than necessary by management. Instead of job completions flowing smoothly into the title-closing stage with customers, there may develop a major title-closing bottleneck. To remedy the problem of unbalanced operations flow in this situation, management must introduce prompt corrective action either through the rescheduling and reallocation of resources or through more effective expediting in problem areas.

Program Variance Control

The program variance control function provides management with a methodical process by which the total variance of overall program performance from established construction goals may be monitored, measured, and controlled. This management process involves:

1. Monitoring overall program progress to ensure that total variance from program schedules is maintained within tolerable limits
2. Measuring the total weekly variance of construction progress from established program performance goals
3. Recording the total weekly variance of the construction program from its schedule timetable and establishing trend series for its continuous evaluation
4. Controlling the total weekly variance of overall construction from the schedule timetable by introducing measures which reduce the level of program variance or by reallocating and rescheduling resources for its elimination

Through the program flow curve technique discussed in the previous section, it is possible for management to make general assessments on the overall effectiveness of the construction program and to initiate action for its improvement. This evaluation technique may be further refined to allow management to make specific assessments for monitoring the overall progress of the program. In order to accomplish this, management must establish specific quantitative values by which the variance between planned and actual construction performance may be measured. The quantitative value used for this measurement is the total number of construction workdays variance between actual and scheduled program progress. The variance value is computed from the job status data posted weekly on the program control chart. It is arrived at by adding all the plus variance figures shown for critical work activities on each job and then subtracting their total minus variance values. The net product of this computation represents the total number of construction work days variance that the overall program is ahead or behind its construction performance goals.

In order to evaluate the significance of each week's construction program variance, determine its trend and possible effect on the overall program, and control its fluctuation within reasonable limits, management must maintain a program variance record as illustrated in Fig. 4.10. The program variance control record provides for weekly entries which indicate the week date of posting and the total number of work-

PROGRAM VARIANCE CONTROL RECORD
FOR CONSTRUCTION WORKDAY VARIANCE FROM SCHEDULE

Week-ending Date	Workday Variance (+)	Workday Variance (−)	Comments
3/5	2		
3/12	3		
3/19	6		
3/25	8		
4/2	7		
4/9	10		
4/14	8		
4/23	8		
4/30	11		
5/7	10		
5/14	13		
5/21	17		
5/28	19		
6/7	25		

Figure 4.10

days variance for the program that week and for management comments relating to the variance information posted.

The first variance entry made in the illustration is for the work week ending 3/5, which was the first week of operations for the construction program. The total program variance for that week, during which two jobs were started, was +2 workdays. By 6/7, the latest entry date on the record, the program variance had increased to +25 workdays. On that date, thirty of the fifty jobs planned for the construction program had been started, with one job already in the title-closing stage. Of the thirty jobs, thirteen were ahead of schedule, seven were behind, and ten were on schedule. At this particular point in time, close to 40 percent of the total calendar period (nine months) planned for the construction program had elapsed, and overall program performance was well ahead of that anticipated at the outset.

The importance of the program variance control technique is that it permits management to summarize in capsule form the overall effectiveness of the entire construction program. The weekly variance data posted on the program variance control record provides a realistic measure of how well management is actually coordinating and controlling the total program. Fluctuations in the level of weekly program variance reflect the overall stability of the program and provide a basis on which management may project whether it will achieve its end goals as scheduled. The program variance trend furnishes management with a directional guide with which decisions may be reached on whether to reallocate resources, reschedule operations, or revise flow plans so that overall objectives may be efficiently realized.

Part II

ORGANIZATION DEVELOPMENT AND CONTROL

Home building is not only the oldest, the largest, and the most geographically diversified industry in America, it is also the most difficult for the development and growth of small business organizations.

In size, the home building industry dwarfs the nation's major manufacturing industries. Its annual production dollar volume is three times that of the automotive industry and more than double that of the petroleum refining and aircraft industries. There are more firms engaged in the construction of homes than there are companies in the automotive, petroleum refining, aircraft, chemical, and rubber industries combined.

The twenty largest companies in each of these major manufacturing industries account for between 80 and 90 percent of their industry's total production output. In sharp contrast, no more than 1 percent of the country's building firms have an annual construction volume of more than 500 homes. Almost two-thirds of all home building firms construct less than 25 homes a year. Whereas the major share of the nation's manufacturing output stems from a comparatively small group of large companies, the bulk of its home building output flows from an extremely broad base of small construction firms.

The obstacles to growth which restrict the expansion of small home building organizations are unlike those confronted by firms in manufacturing industries. Because of their lack of specific management techniques to overcome these obstacles, most small home builders have been unable to break through their low levels of construction volume. Their most serious problems are organization development and operations control, and their business growth is restricted primarily because of the following restraints:

1. Organization restraints—the extremely transient nature of their home building business, their lack of fixed operating facilities, their short-run periods of operations at each construction site, and the intense competition generated by new firms continuously entering the field
2. Operations restraints—the instability and inefficiency of operations resulting from their continuous relocation to new construction sites, their difficulty in establishing and maintaining long-range operations plans, their need for technological improvements in construction methods, and the conflicting building codes and regulations to be complied with at each new construction site.
3. Product restraints—the long life cycle of housing units, their complex physical configuration for production, and the lack of technological advances in the concepts of human shelter
4. Management restraints—the lack of management systems and tools designed specifically for home builder organizations, the need for the development and dissemination of new builder management concepts and techniques, and, most frequently, the lack of an organization environment conducive to the formulation of more scientific management methods

The management control techniques developed for the solution of organization development and operations control problems in other industries cannot be effectively used by home builders in their present form. However, the management concepts upon which these techniques are based apply with equal force and value to firms in all industries. They provide a residual of scientific management principles which can be translated into powerful management control techniques for home builders. Their specific adaptation for the solution of home builder operations control problems has already been demonstrated.

The basic tools employed by successful management entrepreneurs in all industries for the development and control of their organizations are as follows:

1. Organization structures—the structural framework within which the goals of management are translated into a functional organization format for their achievement
2. Delegation and control—the process through which management defines, documents, delegates, and controls the functional duties and responsibilities of all members of the organization in relation to the achievement of its overall mission
3. Systems and procedures—the controls through which management monitors the flow of all administrative performance throughout the organization and with which it is administered as a totally integrated management system

Each of these basic management tools—organization structures, delegation and control, and systems and procedures—is reviewed in terms of its specific application to home builder organizations in the remaining sections of the book. The fundamentals of organization structures and delegation and control are described and demonstrated in detail in the chapters that immediately follow. The management control techniques presented clearly illustrate the organization development and control *modus operandi* for the successful growth of home builder organizations.

CHAPTER 5

Organization for Growth

FUNCTIONAL STRUCTURES

At their outset, most small home building enterprises are informal organizations in which the need for formal structures, job descriptions, and procedures is not essential for business survival. As the building enterprise grows, formal delegation of authority and responsibility becomes more important. Not only must the organization structure expand with an increase in business volume, but the basic control methods used must undergo a change to meet conditions which can no longer be managed by personal contact. Many a successful small home builder has failed when he assumed larger responsibilities and could no longer direct the functions of the enterprise in person. He had not learned the management techniques of organization development and control.

The procedures followed for the construction of a home and the development of a home building organization are quite similar in principle. They involve the formulation of structural plans within which their end goals may be realized, the definition of specific structural elements that will be required and the functions that they will serve, and the establishment of practical working procedures for the accomplishment of their structural goals. To construct a house, a home builder must have:

Organization for Growth 91

1. Architectural plans which schematically formulate the functional relationships and dependencies of all physical components required for its construction
2. Detailed specifications which define and describe the physical characteristics and composition of each of its components
3. Work methods and procedures which determine how the construction of the house may be best carried out in accordance with its plans and specifications

The essential ingredients required for the construction of a house are plans, specifications, and technical know-how. Without any one of these basic ingredients, the construction of a house would be not only ill-conceived, but impractical as well.

Developing a successful home building organization is no less logical and methodical a process than that followed for the construction of a house. The management methodology applied in developing the organization is as disciplined and proceduralized as that followed in the technical structuralization of the finished house. It involves:

1. Formulating a structure of organization through which the functional relationships of all elements involved in its operation may be coordinated and integrated into a total scheme of organization
2. Defining the specific performance requirements of each functional element in the organization and delegating duties and responsibilities to each
3. Establishing systems and procedures which describe in detail how all elements of the organization are to function in relation to each other so that overall organization goals may be efficiently realized

The structuring of the organization, defining of duties and responsibilities, and establishing of operations systems and procedures are indispensable to the development and control of a home building organization. Each of these organizational requirements must be approached and formulated by the builder with a methodology and skill similar to that required for the technical design and construction of his finished products. While a high level of technical knowledge is required for the actual construction of the houses, a comparable, if not greater, level of management knowledge is needed for the development and control of the organization structure responsible for their construction.

The most basic ingredient in the development of an organization is its structural framework. The structure of the organization represents a logical management scheme for translating its objectives into functional elements for their accomplishment. It provides the home builder with a practical basis for:

1. Defining each functional element (department and activity) that is an essential part of the organization
2. Interfacing and controlling each of these functional elements
3. Delegating duties and responsibilities for the management of each functional element
4. Developing systems and procedures for the integration of all activities in each functional element into a total system of organization

The structural framework of the organization must fully reflect the functional characteristics and requirements of the enterprise for which it is designed. It must take into consideration the various levels of organization structure that will be needed for its implementation. Each functional element required for the achievement of its organization objectives must be identified—executive functions, staff functions, departmental functions, and such other basic functions as may be needed by the enterprise. Each functional element is an integral part of the whole organization, and the neglect or omission of but one essential part makes impossible the effective development and control of the organization as a whole. The significant relationship of each functional element in a medium-sized home builder organization is illustrated in Fig. 5.1.

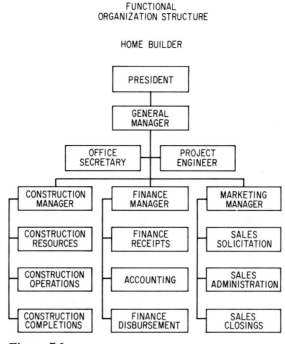

Figure 5.1

Organization for Growth

At the top of the organization is the president and general manager. Beneath the general manager are the secretarial and project engineering staff functions. Below are the department managers to whom the general manager delegates both authority and responsibility for the conduct of their respective departments. The general manager may delegate complete authority and responsibility for the departments, or he may retain certain phases for his function. Whether the department managers have complete or partial responsibility and authority, it is the same as that delegated to the general manager but more limited in scope. The department managers in turn have their subordinate personnel levels, each receiving authority and responsibility for their functions. Management directives move in a direct vertical line from the general manager to subordinates, and the reports which result from such directives are returned back up through the same channels. There is no question of authority or responsibility in the functional organization structure, as each functional element is a complete unit in itself.

The functional organization chart provides the home builder with a valuable visual tool for developing and controlling his overall scheme of organization. It provides a plane of reference for all management in the organization as it:

1. Establishes the chain of command in the organization, from the top position down to the lowest functional element in its structure
2. Identifies each significant functional element in the organization—engineering, finance, construction, and marketing—as an essential part of the whole
3. Illustrates the functional relationships of all management in the organization—the president, general manager, project engineer, construction manager, finance manager, and marketing manager
4. Clarifies the levels of authority and responsibility in the organization: top management—president, executive management—general manager, staff management—project engineer, and department management—finance, construction, and marketing managers

Most important, the functional organization chart provides the home builder with a logical method for integrating all required management resources into a structural plan for the accomplishment of organization goals (see Fig. 5.2).

There are basic conditions which all home builder organization structures must satisfy if they are to successfully achieve their overall goals. They must be simple, easily understood by the management team, and fix their levels of authority and responsibility. They must be capable of adjusting quickly to changing business conditions and pressures which lead to either organization contraction or expansion.

They must provide the format for department and staff management advancement to higher levels of authority and responsibility consistent with the expansion needs of the organization. And finally, they must be soundly balanced in the sense that their structural frameworks are designed to meet both the present organization needs of the enterprise and its long range growth objectives.

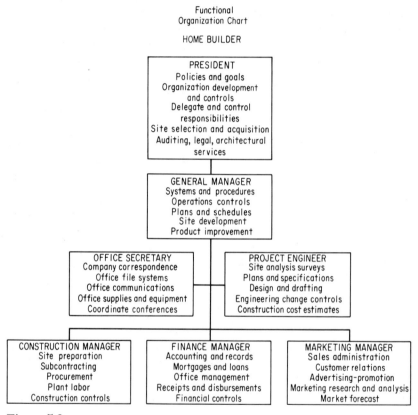

Figure 5.2

In summary there is no ideal nor typical home builder organization structure, as they vary according to the specific needs of individual home builders. While their functional structures may differ in format, their design principles are similar. What is important for success in organization development is patterning the functional elements of the organization so that they best match the organization goals of the enterprise.

ECONOMIC CHARACTERISTICS

The basic goal of a home building enterprise is economic in that it is conceived as an organized means by which financial investments may be made and, in turn, profits may be realized.

This fundamental concept underlies the formation and development of all business organizations. The structure of the business enterprise must be designed to facilitate the achievement of its economic objectives. As an economic entity, the enterprise is totally dependent for its survival and success upon the economic soundness of its organization structure.

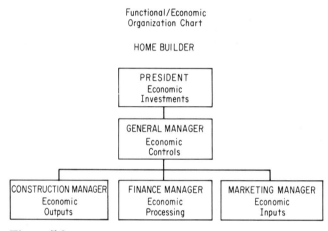

Figure 5.3

With this perspective in mind, home builder management must view the development of its organization structure in terms of the functional elements required for the enterprise to realize its economic goals.

In Fig. 5.3, the major elements of the functional organization structure are graphically illustrated in terms of the economic purposes which they serve:

1. The president—has the responsibility for providing the economic investments necessary for the total organization to function.
2. The general manager—exercises an economic control function over the flow of investments into the organization, balances the operation flows which result, and monitors the profit or loss flow back through the organization to investors.
3. The marketing manager—provides the economic input flow of business from customers for subsequent processing and conversion into output flows created by other major elements of the organization.

4. The construction manager—is responsible for the economic conversion of marketing inputs into the output flow of finished housing units to customers.
5. The finance manager—is accountable for the economic processing of input and output flows, their conversion into deficit or surplus flows, and their diversion back upstream through the organization to its investors.

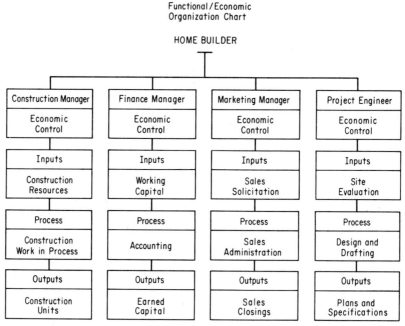

Figure 5.4

When functional organization structures are conceived and designed with the economic goals of the enterprise in mind, the builder can visualize the overall economic scheme of the organization and incorporate into its structure economic characteristics which must be inherent in each major functional element of the organization. These elements may also be viewed in terms of their individual economic structures. Just as the total organization complex is characterized by economic input, output, process, and control functions, the individual major elements of the organization possess similar characteristics, as the chart shown in Fig. 5.4 reveals.

Building organization structures with their economic purposes and characteristics in mind not only simplifies the structural task of organization development, but also ensures that functionally balanced and economically sound structural formats of organization will result. In addition, it both facilitates and simplifies the task of defining and delegating duties and responsibilities to members of the management team and, in turn, permits more effective evaluation of management performance.

FORMAT FOR STRUCTURAL GROWTH

The problem of organization development and growth for the average home building firm is much more complex than that for manufacturing enterprises because of:

1. Their continuous relocation to new construction sites along with geographic movements in the local demand for housing
2. Their comparatively short-run periods of operation at each construction site
3. Their lack of flexible organization structures to accommodate multisite construction operations

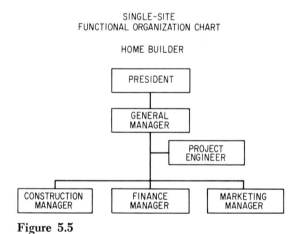

Figure 5.5

The constant relocating of home builders' operations along with geographic shifts in the demand for housing is characteristic of the home building industry. However, the lack of organization flexibility to expand at the local housing level is a limitation that is inherent in the management of the individual home building firm. It therefore follows that flexibility is the most essential growth ingredient in a home builder's organization for structural expansion through multisite construction operations.

The distinct advantage of the functional organization structure, discussed previously and shown in Fig. 5.5, is its flexibility for organizational growth. Its format, in addition to being functional in character to best satisfy the business goals of the organization, is also designed to serve as an economic model in order to effectively meet the expansion needs of the enterprise. It is modular in that its basic format represents a complete economic unit capable of fulfilling the organization control requirements of a single-site construction program (Fig. 5.5) and also of lending itself to structural reproduction for multisite construction operations as illustrated in Fig. 5.6.

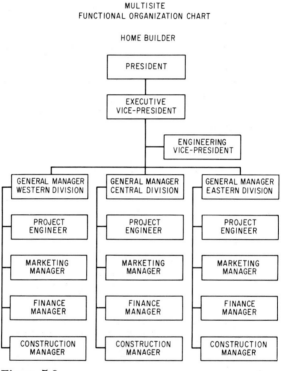

Figure 5.6

In principle, the functional organization structure for a multisite home building firm operates in a manner similar to that for a single-site firm. To accommodate an increase in construction volume, the basic structure of the organization is expanded, modular-like, for the control of each additional construction site or division activated. The levels of authority and responsibility at the higher tiers of management are increased proportionately with the breadth of expansion of the firm's multisite construction operations. Each construction site, if its operations volume warrants, functions as a separate organization module with its own staff and functional management tiers. If construction operations are geo-

Organization for Growth

graphically regionalized, a number of local construction sites may be incorporated under a divisional scheme of organization as shown in Fig. 5.6.

In concept, the entire corporate and divisional organization is a total economic complex, or business model, and each of its divisional organizations is a replica of the total scheme of organization. At the divisional level, new construction site operations are literally plugged in as they are activated and plugged out as they are completed. The functional organization structure provides the much-needed organization flexibility required by the average home builder for business expansion. It may well provide the key to organization development and growth for many home building firms.

CHAPTER 6

Delegation and Control

JOB DESCRIPTIONS

Organization structures represent management schemes for effectively integrating all capabilities and resources in an enterprise into an efficiently organized plan for the accomplishment of its overall objectives. If the scheme of organization is to be successfully implemented, it must be supported with detailed specifications in the form of written job descriptions which pinpoint the specific duties and responsibilities of all key members of the organization. Job descriptions provide management with a practical administrative tool for:

1. Defining and delegating personnel duties and responsibilities
2. Establishing job specifications for the guidance of personnel in the performance of their work activities
3. Developing job standards against which the performance of personnel may be measured
4. Determining the strengths and weaknesses of personnel and their capacity for promotion and growth
5. Developing manpower plans for accomplishing company goals
6. Evaluating the qualifications of applicants for employment

The preparation of job descriptions is a critical step in the organization development and control process. It is the means through which management translates its plan of organization into written specifications which prescribe who in the enterprise is to carry out what specific functional activities to achieve the objectives of the plan. It closes the communications void between the functional objectives inherent in the structure of organization and the functional activities which must be performed for their accomplishment.

Job descriptions provide the formal medium through which top management delegates authority to members of the organization to carry out its objectives. Delegation is the act of assigning functional duties to be discharged by subordinates in the management team, and authority is the responsibility granted subordinates by top management to carry out their functional duties as delegated.

Job descriptions outline in detail the specific duties and responsibilities of each member of the management team. The management team consists of those members of the organization who play a vital role in the administration of the enterprise. In a medium-sized home building organization this may include the president, general manager, project engineer, construction manager, finance manager, marketing manager, and office secretary. Each of these team members contributes significantly in his day-to-day activities toward the efficiency with which the organization is administered.

Job descriptions also provide the bases upon which the organization's communications system is developed. They furnish the medium through which policies and directives are transmitted from top management down through the organization network to the lowest tier in its structure. Through the same channels of communication, the realities and results of day-to-day activities are transmitted back upstream to top management in the form of records and reports. Thus, job descriptions provide both the activation medium for communicating duties and responsibilities down through the organization and the reporting medium for transmitting results back up through the organization to top management.

To be effective, job descriptions must reflect the specific functional performance requirements of each member of the management team. They must describe in detail the particular functional element of the organization for which each member of management is responsible. They must provide the functional guidance necessary to enable each individual to carry out his specific duties and responsibilities in the most effective manner possible.

Presented in the remaining portion of this chapter are job descriptions which illustrate those that would be needed by a medium-sized home building firm. They demonstrate the format and contents of job descriptions for each key member of the organization's management team and are patterned after the functional organization structure shown in Fig. 5.2. In addition, they are sufficiently detailed in their contents to serve as a basis for the guidance of home builders in the preparation of similar job descriptions for their management teams.

PRESIDENT

DUTIES AND RESPONSIBILITIES

The president is responsible for:

1. Establishment of company policy
2. Development of the company's organization structure
3. Definition of management functions and delegation of duties and responsibilities
4. Establishment of company objectives
5. Determination of administrative systems and procedures required for company management
6. Acquisition of raw land for development and construction
7. Development of a respected company image
8. Establishment of sound management practices
9. Realization of a favorable rate of return on the company's investments
10. Selection of markets for company products
11. Acquisition of professional services to assist in company management
12. Acquisition of capital and investments for company growth

GENERAL MANAGER

DUTIES AND RESPONSIBILITIES

General Duties

The general manager is directly responsible to the president for the development of company management functions, implementation of administrative systems and procedures, and management of its land development, engineering services, product design, planning and scheduling, and program control activities.

Specific Responsibilities

MANAGEMENT FUNCTIONS

1. Acquire capable management personnel to supervise departmental operations.
2. Prepare job descriptions outlining the duties and responsibilities of each member of management.

Delegation and Control 103

3. Monitor the performance of each department manager to ensure that the company's operating goals are realized.

SYSTEMS AND PROCEDURES

1. Define the administrative systems and procedures necessary for the efficient performance of departmental functions.
2. Implement administrative systems and procedures for the guidance of personnel in each department.
3. Prepare a systems and procedures manual as a source of reference for the conduct of all administrative matters within and between departments.
4. Monitor the administrative performance of each department to ensure that functions are discharged in accordance with established systems and procedures.

LAND IMPROVEMENT

1. Supervise the preparation of topographical and preliminary maps for proposed streets, lots, and utilities.
2. Represent the company in administrative matters with planning boards, inspection agencies, and engineering sources.
3. Obtain necessary improvement permits from local government authorities.
4. Arrange for land clearance preparatory to improvement operations.
5. Obtain bids and award contracts for land improvements.
6. Schedule and supervise all land improvement activities.
7. Coordinate land improvement activities with construction operations.

OPERATIONS PLANNING AND SCHEDULING

1. Establish planning and scheduling systems to control land development and construction operations.
2. Maintain an operations control center to devise, develop, implement, and monitor the planning and scheduling of all land development and construction operations.
3. Conduct weekly meetings with all members of construction management to review the status of land development and construction operations.

ENGINEERING SERVICES

Acquire such professional engineering services as may be necessary to accomplish land development and construction objectives.

MANAGEMENT CONTROLS

1. Establish management control measures (records, schedules, reports) to monitor the functional performance of each department.
2. Implement management evaluation techniques (charts, schedules, reports) to monitor the integrated performance of all departments in the company.
3. Conduct weekly executive committee meetings with department heads to review operations in each department, and introduce management recommendations to improve overall company performance.

APPROVED BY: _____

DATE: _____

PROJECT ENGINEER

DUTIES AND RESPONSIBILITIES

General Duties

The project engineer is directly responsible to the general manager for the preparation of site analysis surveys, design and drafting, project plans and specifications, and construction cost estimates.

Specific Responsibilities

PLANS AND SPECIFICATIONS
1. Supervise engineering analysis and release of plans and specifications.
2. Maintain plans and specifications register for all engineering documents.
3. Establish engineering coordination between company, architect, and engineering firms.
4. Maintain up-to-date engineering files for all project plans and specifications.
5. Supervise engineering review and approval of all plan and specification modifications.
6. Furnish subcontractors and suppliers with plans and specifications as required.

DESIGN AND DRAFTING
1. Provide design and drafting services for company product improvements and customer changes.
2. Coordinate the flow of construction change orders with marketing, finance, and construction personnel to ensure their processing in a timely and efficient manner.
3. Establish a change order control system to monitor the processing of construction changes so that they do not delay field operations.

SITE ANALYSIS SURVEYS

Conduct site analysis surveys for the purpose of evaluating the economic, architectural, and engineering characteristics of prospective construction sites.

COST ESTIMATES
1. Develop cost estimates for construction jobs from preliminary prints, and prepare final cost estimates on each job before construction is started.
2. Obtain cost estimates from subcontractors and suppliers for all construction activities.
3. Establish budgetary cost controls for each job before the start of construction.
4. Monitor actual job cost against budgeted performance cost, and recommend construction improvements to reduce costs.

APPROVED BY: _____

DATE: _____

FINANCE MANAGER

DUTIES AND RESPONSIBILITIES

General Duties

The finance manager is directly responsible to the general manager for the negotiation and administration of mortgages and loans; the establishment of accounting systems and record-keeping procedures; the implementation of expense, budget, wage, and salary controls; and the management of the main office and personnel programs.

Specific Responsibilities

MORTGAGES AND LOANS

1. Establish procedures for satisfying company financial requirements through business loans, suppliers' and subcontractors' credit, construction loans, and such other sources as are deemed necessary.
2. Establish procedures to maintain a sound financial balance between capital investment and indebtedness.
3. Establish ways and means to increase the efficiency of working capital and to reduce indebtedness to a practical minimum.

ACCOUNTING AND RECORD KEEPING

1. Establish an accounting system with supporting records to monitor and safeguard the financial position of the company.
2. Prepare monthly operating statements showing the current financial status of the company.
3. Prepare *pro forma* operating statements projecting the anticipated future earnings and financial position of the company.
4. Establish accounting systems for the documentation of cash receipts and disbursements in journals, ledgers, earnings and cost records, and financial statements.
5. Establish control measures to monitor the cash and profit condition of the company.

BUDGET CONTROLS

1. Prepare quarterly and annual operational budgets based on construction expense estimates, overhead estimates, and sales forecasts.
2. Establish monthly direct expense budgets for anticipated construction costs.
3. Establish monthly indirect expense budgets for anticipated administrative, general, and selling costs.
4. Establish gross and net profit goals based on budgeted direct and indirect costs.
5. Establish a working capital budget based on budgeted direct and indirect costs.
6. Establish control measures to monitor the budget program.

OFFICE MANAGEMENT

1. Establish employee benefits and service programs as pertains to insurance, sick leave, holidays, and vacations.

2. Supervise the office staff, and establish employee procedures for the discharge of administrative and clerical duties.
3. Establish work schedules, hours of duties, lunch periods, and overtime procedures.
4. Establish a schedule of wages and salaries for all hourly and salaried personnel.

Relations with Other Departments

1. Advise on the prevailing level of wages and salaries paid to construction-industry employees in the region.
2. Advise on company policies regarding employee benefits and services.
3. Provide centralized clerical and typing services as required by other departments.
4. Provide department heads with monthly budgetary performance reports.
5. Attend weekly executive committee meeting.

APPROVED BY: _____

DATE: _____

CONSTRUCTION MANAGER

DUTIES AND RESPONSIBILITIES

General Duties

The construction manager is directly responsible to the general manager for the management of all construction operations, procurement and subcontractor source selection and negotiations, equipment maintenance and usage, customer services, and program safety measures.

Specific Responsibilities

CONSTRUCTION

1. Approve all construction plans and specifications prior to construction starts.
2. Make recommendations on basic construction techniques.
3. Maintain and monitor production controls and schedules.
4. Determine requirements for plant personnel and mechanical equipment.
5. Supervise and schedule work requirements for plant personnel.
6. Conduct daily field inspections of all work under construction.
7. Catalog and recommend usage of new products and materials to improve product quality.
8. Prepare housekeeping rules for plant labor to maintain a clean construction site at all times.
9. Prepare job closing inspection reports for completed construction.
10. Attend weekly executive committee meetings to coordinate construction activities with the work of other departments.
11. Approve all subcontractor and supplier invoices making sure that each is backed by subcontractor agreements and purchase orders.

Delegation and Control

12. Review vendor and subcontractor billing procedures, and recommend changes to facilitate quick handling, discounting, approval, and voucher preparation.
13. Approve all utility and telephone bills.
14. Supervise quality control inspection of subcontractor services and supplier materials prior to approval of invoices for payment.
15. Establish inventory control methods and procedures to ensure availability of critical materials.
16. Establish procurement control procedures for the smooth flow of purchase orders.

EQUIPMENT

1. Determine the need and timing for the lease and procurement of all construction equipment.
2. Negotiate the conditions of lease and procurement for all equipment.
3. Maintain an equipment register with a descriptive listing of all equipment on hand, indicating date of acquisition or lease, present location, and general operating condition.
4. Prepare monthly cost analysis reports on the maintenance and usage of equipment.

PROCUREMENT AND SUBCONTRACTING

1. Furnish each subcontractor and supplier with written specifications for work and material requirements.
2. Obtain subcontractor and supplier prices for construction work and supplies.
3. Prepare lumber lists, brick lists, etc., in adequate time to ensure that construction activities are not delayed.
4. Negotiate price, performance, and delivery for materials and services with suppliers and subcontractors prior to the start of construction.
5. Work with subcontractors and suppliers on problem areas to ensure that construction is not delayed.
6. Supervise the hiring and firing of all subcontractors, suppliers, and plant labor.
7. Prepare a monthly directory of subcontractors and suppliers, and distribute it to all department managers.
8. Administer all contracts with subcontractors ensuring that their performance is in accord with the terms of subcontractor agreements.
9. Inspect all incoming materials to ensure that quantity and quality is as specified in purchase orders.

SAFETY PROGRAM

1. Maintain safety regulations at the construction site.
2. Coordinate security patrols for the protection of company property.
3. Coordinate all matters pertaining to compensation insurance for labor, and take necessary steps to maintain lowest possible rate.

CUSTOMER SERVICES

1. Resolve all customer complaints during construction.
2. Prepare and furnish price quotations on construction changes and special items.

Relations with Other Departments

MARKETING DEPARTMENT
1. Process requests for construction changes.
2. Process customer selections reports.
3. Furnish information on house completion schedules.
4. Furnish information on action taken on customer complaints.
5. Quote customer prices for construction changes and special items.

FINANCE DEPARTMENT
1. Furnish information required for accounting records.
2. Furnish information on inventories of materials, work in progress, and construction equipment.
3. Furnish construction schedules and information on completion of houses.
4. Furnish information and assistance on plant payroll matters.

APPROVED BY: _____

DATE: _____

MARKETING MANAGER

DUTIES AND RESPONSIBILITIES

General Duties

The marketing manager is directly responsible to the general manager for sales administration, sales promotion, sales planning and forecasting, advertising, market research and analysis, and customer relations.

Specific Responsibilities

SALES ADMINISTRATION
1. Organize and manage the sales department to ensure that sales goals are realized.
2. Implement administrative practices and procedures for the preparation and processing of sales memoranda, agreements, and settlements.
3. Prepare and issue a sales manual to guide personnel in the handling of all customer contacts.
4. Conduct weekly sales conferences to review and improve sales performance.
5. Attend weekly executive committee meetings to coordinate marketing activities with the work of other departments.

SALES PROMOTION
1. Develop and implement a publicity release program through newspapers and magazines.
2. Develop a public relations program to create a favorable company image in the community.
3. Establish a display center to exhibit materials used for construction, and relate the location of the site to community conveniences.

Delegation and Control

PLANNING AND FORECASTING
1. Develop sales forecasts and plans for house sales for the span of the construction program.
2. Present sales plans to other departments for their planning and scheduling requirements.
3. Evaluate sales performance against the sales plan, and introduce measures to ensure that sales goals are reached.

ADVERTISING
1. Develop an advertising program with expenditure budgets and schedules for local media.
2. Prepare sales brochures and literature for use with sales promotion.
3. Coordinate the advertising and sales promotion programs.

MARKET RESEARCH AND ANALYSIS
1. Collect marketing data relating to the size of the market, its characteristics, and economic trends.
2. Study attitudes, reactions, and preferences of local consumer groups to ensure that the product design is in line with what future buyers require and will buy.
3. Conduct area surveys to develop general information on market characteristics of the region, such as inventories of unsold new houses, employment and income, population trends, migration and mobility, and the volume and value of new home building.
4. Conduct local surveys to determine general market conditions, such as the number of prospects interested in purchasing new houses, the characteristics of prospective buyers, consumer preferences, buying intentions of prospects, plans and capabilities of potential competitors.

CUSTOMER RELATIONS
1. Develop a code of conduct which ensures that all customer relations are on a high ethical level.
2. Develop a standard procedure for the administration of customer selections.
3. Develop a standard procedure for the effective processing of all customer services.

Relations with Other Departments

CONSTRUCTION DEPARTMENT
1. Furnish construction with written information on customer change requests.
2. Furnish copies of sales agreements and customer selection reports.
3. Furnish updated copies of the sales forecast each month.

FINANCE DEPARTMENT
1. Advise when dates are set for signing sales agreements.
2. Furnish updated copies of the sales forecast each month.
3. Furnish signed copies of all sales documents.
4. Furnish completed copies of the customer selection report.

APPROVED BY: _____

DATE: _____

OFFICE SECRETARY

DUTIES AND RESPONSIBILITIES

General Duties

The office secretary is directly responsible to the general manager for handling company correspondence, maintaining office files, coordinating office communications, providing duplicating services, ordering office supplies, and coordinating company meetings and customer conferences.

Specific Responsibilities

COMPANY CORRESPONDENCE
1. Pickup, open, and sort mail.
2. Take dictation and type company correspondence.
3. Type legal, financial, and other types of forms required in the normal course of company business.

OFFICE RADIO, DUPLICATION, AND TELEPHONE
1. Coordinate the receipt and dispatch of operations information between the construction site and the main office with the radio communications system.
2. Provide duplicating services for the reproduction of correspondence and forms as required in the normal course of business.
3. Place outgoing telephone calls, and answer all incoming calls.

OFFICE FILING
1. Establish central filing systems for the storage of correspondence and records.
2. Organize the structure of the filing system so that correspondence and records are maintained on a subject basis and in chronological order.
3. Implement file control techniques which will provide ready access to, and retrieval of, correspondence and records.

OFFICE SUPPLIES
1. Order stationery, forms, and sundry office supplies as required for office administration.
2. Inventory office supplies to ensure that adequate sundries are available to satisfy normal administrative needs.

APPOINTMENT AGENDA
1. Maintain appointment register to record and schedule company meetings.
2. Coordinate appointments and conferences to ensure that business meetings take place as scheduled.

APPROVED BY: _____

DATE: _____

Part III

SYSTEMS AND PROCEDURES

There are a number of effective management techniques through which home builders may coordinate and control the administrative performance of their organizations. The most powerful of these management control tools are:

1. Functional organization charts, which schematically integrate and control the structural relationships of all functional elements in the home builder's organization
2. Job descriptions of key members of the organization's management team, through which the home builder may delegate and control their functional performance in relation to the organization's goals
3. Systems and procedures, through which the home builder methodizes and controls the flow of administrative activities and information both within and through each functional element of the organization

Functional organization charts and personnel job descriptions are basic control techniques of management and have already been reviewed. Systems and procedures, which will be described in this section, provide the home builder with a third management control technique, as they virtually lace the overall organization complex into a total scheme of administrative management.

Each of these management control techniques contributes in significant measure to the overall coordination of the home builder's organization. They are intricately related to one another, and the effective utilization of one is dependent upon the proper implementation of the others. Their relationships and dependencies are described as follows:

1. The organization chart, the most basic tool, provides a structural framework within which all the integral parts of the organization may be unified into a total complex and thereby coordinated and controlled.
2. Within this framework, job descriptions provide a functional control mechanism by coordinating and controlling the vertical flows of delegation and authority from the policy-making level down through the working levels of the organization.
3. Also within this framework, systems and procedures provide a methodization function, coordinating and controlling the horizontal flows of all activities and information which cut across each functional component of the organization.

Collectively, these three management control tools integrate the structure of the organization, the functional flows of delegation and authority, and the horizontal flows of systems and procedures into a total system of organization.

Systems and procedures are implemented in an organization after its structural and functional controls have been established. They may be defined as follows:

1. Systems represent logical assemblies of related procedural activities, which, when collectively organized, provide management with systematic methods for coordinating and controlling the major areas of administration in an enterprise. In a home builder's organization they may include engineering, finance, construction, purchasing, subcontracting, inspection, marketing, and filing systems.
2. Procedures are logical management processes which outline in sequential form all of the actions which must be taken for the accomplishment of a particular administrative requirement. The procedures required in a home building organization would include, among others, purchasing order

procedures, title-closing procedures, and job work order procedures.

The systems approach to administration is important because it provides management with a logical means for identifying, organizing, and controlling the individual procedural activities associated with each major area of administration in an enterprise. It not only methodizes and simplifies the overall flow of administrative activities and information through the organization, but greatly improves the overall efficiency of the enterprise as well. To illustrate, the purchasing system in a home building organization incorporates each important administrative procedure related to its functional performance:

Purchasing System

1. Request for quotation procedure
2. Purchase order procedure
3. Purchase order register procedure
4. Vendor certification procedure
5. Vendor register procedure

This home builder purchasing system takes into consideration each important procedural activity involved in its administration. In the procurement cycle, requests for quotations must be released for material requirements, purchase orders issued to successful bidders to supply the materials, purchase order registers maintained to follow up on the flow of material deliveries, vendor certifications made to approve vendor invoices for payment on materials delivered, and vendor registers maintained to list present and potential supply sources. The purchasing system is designed to coordinate and control all procedures which must be followed for its successful implementation in the home builder's organization.

Procedures are prepared in instruction form to guide personnel in the performance of their administrative duties. Each individual procedure is designed for a particular administrative requirement. The format of the procedure indicates:

1. The management source that authorized its preparation
2. The personnel involved in its implementation
3. The subject title of the procedure
4. A description of its purpose
5. The personnel responsible for its implementation

6. Any forms that may be required for its implementation
7. The procedural steps to be taken for its implementation
8. The signature of the person authorizing the procedure and the date of approval

Presented in the remaining text of the book are detailed engineering, finance, construction, purchasing, subcontracting, inspection, marketing, and filing systems and procedures. They are designed to guide home builders in the preparation of similar systems and procedures for controlling the overall flow of administration in their organizations.

CHAPTER 7

Engineering Procedures

ENGINEERING PROCEDURE

FROM: General Manager
TO: Project Engineer
SUBJECT: Site Analysis Surveys

Purpose
The purpose of this procedure is to establish a routine method for the preparation and distribution of site analysis survey reports made prior to the acquisition of land for new construction projects.

Responsibility
The project engineer is responsible for conducting site analysis surveys for prospective construction sites as directed by the general manager and for the subsequent preparation of reports related thereto.

Survey Report Format
To ensure that the coverage and depth of the site analysis survey provide an adequate basis for management decisions, it is essential that the survey findings include the following information:

A. *Engineering*
   1. Site features
      a. General description
   2. Foundations
      a. Test borings
      b. Subsurface investigations
   3. Utilities
      a. Storm sewage
      b. Sanitary sewage
      c. Water mains
      d. Electricity distribution
      e. Gas distribution
B. *Architectural*
   1. Building codes
   2. Zoning, setbacks, buffers, etc.
   3. Licenses and fees
   4. Topographical surveys
   5. Property line surveys
   6. Aerial surveys
C. *Pricing*
   1. Site work
   2. Concreting
   3. Masonry
   4. Roofing
   5. Plumbing
   6. Heating
   7. Lighting
   8. Carpentry
   9. Plastering
D. *Real estate*
   1. Site location
   2. Access—traffic
   3. Area buying power
   4. Competition
E. *Summary*
   1. Conclusions
   2. Recommendations

Procedure
1. Site analysis surveys for prospective construction projects are conducted by the project engineer as directed by the general manager.
2. Survey findings are summarized in the recommended report form by the project engineer, and copies are distributed as follows:
   a. One copy is forwarded to the general manager.
   b. One copy is forwarded to the office secretary for file in the survey report file.
   c. One copy is retained by the project engineer.

3. The general manager's copy of the site analysis survey report provides the basis for recommending the approval or disapproval of a new construction site to the president.

APPROVED BY: _____

DATE: _____

ENGINEERING PROCEDURE

FROM: General Manager
TO: Project Engineer
SUBJECT: Project Plans and Specifications

Purpose
The purpose of this procedure is to establish a routine method for processing and distributing project plans and specifications received from architects.

Responsibility
The project engineer is responsible for reviewing, approving, and distributing project plans and specifications.

Procedure
1. Following their preparation, five sets of the following project plans and specifications are forwarded to the general manager by the architects:
   a. Site drawings and specifications
   b. Architectural drawings and specifications
   c. Electrical drawings and specifications
   d. HVAC drawings and specifications
   e. Plumbing drawings and specifications
2. The project engineer is furnished one set of the project plans and specifications for a technical analysis of their completeness.
3. The project engineer returns the project plans and specifications with approval or disapproval comments to the general manager.
4. If the project plans and specifications are technically approved, they are forwarded by the general manager to the office secretary for the following distribution:
   a. One set is forwarded to the construction manager for field office use.
   b. One set is retained by the office secretary in the project file.
   c. One set is forwarded to the project engineer for his file.
5. The remaining drawing and specification sets are retained in the project file by the office secretary for release to subcontractors and suppliers as directed by the construction manager.

6. If the project engineer's technical analysis reveals that the project plans and specifications are in error, they are returned to the architect by the general manager for correction.

APPROVED BY: _____

DATE: _____

ENGINEERING PROCEDURE

FROM: General Manager
TO: Project Engineer
SUBJECT: Construction Change Orders

Purpose
The purpose of this procedure is to establish a routine method for handling all changes in construction plans and specifications.

Definitions
1. *Change in specifications:* any modification of the original construction design
2. *Change order:* a technical directive issued by the construction manager outlining the detail and scope of design modifications
3. *Company change order:* those design modifications initiated by the company for product improvement
4. *Customer change order:* those design modifications initiated by the customer to alter construction design to meet personal requirements

Responsibility
The project engineer is responsible for the preparation and issuance of all construction change orders and for the subsequent coordination of their implementation into construction.

Change Order Form
Attached is a sample of the construction change order form prescribed for use in this procedure. Completion of the form requires entering the following information in the spaces provided:

1. Date—that the change order was prepared
2. Change order number—the control number assigned this change order
3. Company change—check if company has originated
4. Customer change—check if customer has originated
5. Change price—final price calculated for the change

6. Work order number—the number of the work order issued to implement the change
7. Job number—the number assigned to the house
8. Model number—the model style
9. Project—the construction site for the job
10. Location—the off-site location for the job
11. Customer—the buyer's name
12. Phone—the buyer's telephone number
13. Address—the buyer's present address
14. Description of change—a technical write-up of the change (attach additional sheets and drawings if necessary)
15. Approvals—the company processing and customer acceptance approvals and dates

Procedure
A. *Company change orders*
  1. The general manager outlines the scope of the design change for the project engineer and directs that a change order form be prepared and issued.
  2. The project engineer completes the change order form, describing in detail the scope of the construction change and preparing change drawings and specifications, signs the change order form, and forwards change order documents to the construction manager.
  3. The construction manager prepares a cost estimate statement for the labor and material required for the modification, signs the change order form, and forwards it with change order documents to the finance manager.
  4. Finance applies overhead and profit rates to the change order costs, adjusts the established house price, signs the change order form, and forwards the documents to the project engineer.
  5. The priced-out change order is forwarded with supporting documents to the general manager for approval.
  6. The general manager reviews the change order, signs his approval, and returns the form to the project engineer for release.
  7. The project engineer reproduces revised drawings and specifications and forwards them with a copy of the change order to the construction manager.
  8. Construction manager prepares job work orders and forwards them with revised drawings and specifications to subcontractors and suppliers. (The job work order procedure is followed to implement the change order.)
B. *Customer change orders*
  1. Customer change orders are initiated upon the receipt from the customer of a written quotation request for a design change by marketing.
  2. The request for the design change is forwarded by marketing to the general manager.

3. The general manager approves the request and forwards the design change to the project engineer for preparation of a change order.
4. The project engineer completes the change order form, describing in detail the scope of the design change and preparing change drawings and specifications, signs the change order form, and forwards change documents to the construction manager.
5. The construction manager prepares a cost estimate statement for the labor and material required for the modification, signs the change order form, and forwards it with change documents to the finance manager.
6. Finance applies overhead and profit rates to the change order form and forwards the change documents to the general manager for approval.
7. The general manager approves the change order price, signs the change order form, and forwards the change documents to marketing for customer acceptance.
8. Marketing obtains the customer's endorsement for the change order and forwards the change documents to the project engineer for release.
9. The project engineer reproduces the revised drawings and specifications and forwards them with a copy of the change order to the construction manager.
    A copy of the change order form is also forwarded to the office secretary for file in the job folder.
10. The construction manager forwards the revised drawings and specifications with job work orders to subcontractors and suppliers.
    (The job work order procedure is followed to implement the change order.)

APPROVED BY: _____

DATE: _____

CONSTRUCTION CHANGE ORDER

Job No._____ Model No._____ Date_____
Project_____ Change Order No._____
Location_____ Company_____ Customer____
Customer_____ Phone_____ Change Price_____
Address_____ Work Order No._____

Description of Change

_____
_____
_____
_____
_____
_____
_____
_____
_____
_____
_____
_____
_____
_____

Approval   *Date*   *Approval*   *Date*
Marketing Manager_____   Finance Manager_____
Project Engineer_____   General Manager_____
Construction Manager_____   Customer_____

ENGINEERING PROCEDURE

FROM: General Manager
TO: Project Engineer
SUBJECT: Change Order Control

Purpose
The purpose of this procedure is to establish an effective means for monitoring and controlling the processing of construction change orders.

Responsibility
The project engineer is responsible for establishing and maintaining control over all change orders proposed and processed for construction.

Control Form
Attached is a sample of the change order control form to be used as prescribed in this procedure. The form makes provision for posting the following information:

1. Change order number—the number of the change order
2. Date prepared—the date of issue
3. Job number—the number of the job to which it applies
4. Description—a brief description of the change
5. Approval dates—when approved by the project engineer, general manager, and customer
6. Comments—relating to its approval or rejection

Procedure
1. The change order control forms are to be maintained by the project engineer in a file folder, ready for review and discussion with company personnel or the customer as required.
2. Entries are made on the change order control form when change orders are issued, approvals are made, and final disposition comments are posted.
3. The general manager is to be notified weekly of the number of change orders awaiting final approval and the amount of time currently required for their processing cycle.

APPROVED BY: _____

DATE: _____

CHANGE ORDER CONTROL

Change Order No.	Date Prepared	Job No.	Description	Approval Dates			Comments
				Product Engineer	General Manager	Customer	

CHAPTER 8

Finance Procedures

FINANCE PROCEDURE

FROM: General Manager
TO: Finance Manager
SUBJECT: Construction Loan Applications

Purpose
The purpose of this procedure is to establish a routine method for the preparation and submission of applications to lenders for construction loans.

Responsibility
The finance manager is responsible for the preparation and submission of all applications to lenders for construction loans.

Procedure
1. The finance manager obtains a signed construction contract from the customer and/or prepares cost estimates with detailed cost breakdown, names of subcontractors and suppliers with the amount of their bids, and estimated selling price of the house.
2. Financial statements are obtained from the customer, and credit reports are ordered by the finance manager.

3. A property description with plans and specifications is prepared by the general manager in conformance with local building codes, sanitary laws, zoning ordinances, and deed restrictions and is furnished to the finance manager.
4. The construction loan application and a letter of transmittal are prepared by the finance manager for the general manager's signature.
5. The construction loan application with supporting exhibits is forwarded by the finance manager to the company attorney.
6. The company attorney reviews the application and supporting exhibits for completeness and then forwards them to the lender.
7. The lender's appraiser evaluates the lot in respect to the house that is to be constructed.
8. The lender's appraiser determines the local market demand for houses of the proposed design and price in the lot's location.
9. The lender's loan committee reviews the appraiser's findings and approves the construction loan application.
10. The lender forwards to the company a letter of commitment for the requested construction loan.

APPROVED BY: _____

DATE: _____

FINANCE PROCEDURE

FROM: General Manager
TO: Finance Manager
SUBJECT: Construction Loan Closings

Purpose
The purpose of this procedure is to establish a routine method for processing construction loan closings.

Responsibility
The finance manager is responsible for coordinating the preparation and submission of all documents required for construction loan closings.

Procedure
1. Following the receipt of a lender's commitment for a construction loan, the general manager orders a location survey of the lot by the surveyor.
2. The location survey is made by the surveyor and forwarded to the general manager.
3. The general manager forwards the location survey to the finance manager and requests that a title report be made.

4. The finance manager forwards the location survey to the title company and requests the title search.
5. The title company completes the title search and forwards a title report to the finance manager.
6. The location survey and title report are forwarded to the company lawyer by the finance manager with a letter of transmittal suggesting a date for the construction loan closing.
7. The company lawyer forwards the location survey and title report to the lender and establishes a loan closing date.
8. The construction loan closing meeting is held at the lender's office and is attended by the company lawyer, the general manager, and representatives of the lender and the title company.
9. The construction loan closing statement is prepared by the lender indicating:
    a. Closing costs incurred for title search, location survey, credit reports, service charges, recording fees, inspection, appraisal, notary fees, revenue stamps, drawing the papers, closing the loan, etc.
    b. Prepaid expenses for interest, real estate taxes, and hazard insurance
10. The construction loan agreement is signed by the lender and the general manager.
11. The general manager furnishes the lender:
    a. Construction deposits received or held
    b. Evidence of contractor's liability and workmen's compensation insurance coverage
    c. Contract performance bond from the surety company
    d. A preconstruction affidavit that no work has been started on the lot and no material delivered or used
    e. Evidence of hazard insurance policy held
12. After closing, the mortgage is filed for record in the county office.
13. After the mortgage is recorded, the lender makes a preconstruction lot inspection to make sure that no work has been done and no material delivered.
14. The lender then notifies the company in writing to start construction.

APPROVED BY: _____

DATE: _____

FINANCE PROCEDURE

FROM: General Manager
TO: Finance Manager
SUBJECT: Construction Loan Receipts

Purpose
The purpose of this procedure is to establish a receipt schedule for lender

Finance Procedures 127

loan payments that are due as construction work is completed, in accordance with the terms of the construction loan agreement.

Responsibility
The finance manager is responsible for the preparation and submission to the lender of certifications for work completed, as specified in the construction loan agreement.

Construction Loan Receipts Schedule
1. *First payment:* rough enclosure—30 percent of building loan. The building should be framed and sheathed, the finished roof laid and completed, and window and exterior door frames set. Underflooring should be laid, and partition work should be set. Guaranteed survey is submitted showing outside lines and projections to roof (foundation survey).
2. *Second payment:* full enclosure—30 percent of building loan. Exterior finish of brick veneer, siding, shingles, or stucco must be completed and all exterior woodwork primed. Chimney must be built, all rough plumbing and electric BX wiring installed, and stairs up. Lathing should be completed, window sash should be installed, and brown mortar or plaster completed. If sheetrock is used, it must be placed, and all nail holes and joints spackled or plastered. All plastering must be completed. Water and sewer connection to street must be completed, or cesspool connected. Bathtub should be set. Concrete cellar floor should be laid.
3. *Third payment:* fixtures—20 percent of building loan. Tile work must be completed and plumbing fixtures installed. Gutters and leaders must be installed. Heating boiler must be connected, domestic hot-water heating equipment installed, and kitchen cabinets hung. Finish floor should be laid, insulation installed, and all trim completed.
4. *Fourth payment:* final payment—20 percent of building loan. The exterior woodwork should have final coat of paint. Decorating must be completed, linoleum laid, and floors finished. Heating burner, radiators, range hardware, and lighting fixtures must be completely installed. Sidewalks, steps, and terraces must be completed. Grading should be completed, top soil applied, and shrubbery and lawn planted. Curbs should be in and street paved. The building should be entirely completed as called for in plans and specifications and ready for occupancy. All kitchen appliances should be installed. Gas and electric service connections must be made.

Notice of Payments
1. The lender must be given at least forty-eight hours notice of payments due.
2. Interest will be charged on the advances from the date of payment only, on the amount advanced and not on the full amount of the loan from the date of closing.
3. Credit will be given in the percentage of the advance for any additional work completed beyond that listed in the above schedule when inspection is made, except on the final payment which is 20 percent.

Procedure
1. Following the lender's inspection of the first stage of construction (rough enclosure of building), the construction manager obtains two copies of the lender's approved compliance inspection report and forwards them to the finance manager.
2. The finance manager certifies by letter to the lender that the first stage of construction has been completed and requests payment. (The finance manager must give the lender forty-eight hours notice that the payment is due.) Copies of the lender's inspection reports and the letter of certification are forwarded to the office secretary for file in the job folder.
3. The lender forwards to the finance manager the first-stage payment of the construction loan.
4. The finance manager records the loan payment received and deposits the funds in the bank.

(*Note:* The same procedure is followed when certifying for payment of work completed after the second, third, and fourth stages of construction have been finished and approved compliance inspection reports have been received.)

APPROVED BY: _____

DATE: _____

FINANCE PROCEDURE

FROM: General Manager
TO: Finance Manager
SUBJECT: FHA Conditional Commitment Applications

Purpose
The purpose of this procedure is to establish a routine method for the preparation and submission of FHA conditional commitment applications.

Responsibility
The finance manager is responsible for the preparation and submission of all applications for FHA conditional commitments.

Application Form
FHA conditional commitment applications must be prepared and submitted on FHA Form No. 2800 (Rev. 8/65), with the following exhibits:
1. Four plot plans
2. Three grading and location plans

Procedure
1. The finance manager originates the FHA conditional commitment application form with pertinent company information. It is then forwarded to the general manager for completion and preparation of a legal description of the property, four plot plans, and three grading and location plans.
2. The general manager completes the technical sections of the application form and returns it to the finance manager with the requested legal description, plot plans, and grading and location plans.
3. The FHA application form is completed by the finance manager and returned to the general manager for signature.
4. The general manager checks the application form for completeness, signs it, and returns it to the finance manager.
5. The FHA application with exhibits is then forwarded by the finance manager to the mortgagee for review and approval.
6. After approval, the mortgagee forwards the application and exhibits to the FHA office for review and acceptance.
7. The FHA office approves the application and forwards a conditional commitment to the mortgagee.
8. The conditional commitment is approved by the mortgagee and forwarded to the finance manager.
9. Finance makes duplicate copies of the conditional commitment and distributes them as follows:
    a. The original copy is forwarded to the office secretary for file in the job folder.
    b. One copy is forwarded to the marketing manager for use in advising customers that the conditional commitment has been received from the FHA.
    c. One copy is retained by the finance manager for use in certifying to the mortgagee that work has been performed and requesting payments therefor.

APPROVED BY: _____

DATE: _____

FINANCE PROCEDURE

FROM: General Manager
TO: Finance Manager
SUBJECT: FHA Firm Commitment Applications

Purpose
The purpose of this procedure is to establish a routine method for the preparation and submission of FHA firm commitment applications.

Responsibility

The finance manager is responsible for the preparation and submission of all applications for FHA firm commitments.

Application Form

FHA firm commitment applications must be prepared and submitted on FHA Form No. 2900 and accompanied by the following exhibits:
1. Three copies of the contract of sale
2. Three copies of customer's verification of employment forms
3. Three copies of customer's verification of bank deposit forms
4. Three sets of customer's financial statements
5. Three customer credit references
6. Four customer credit reports

Procedure
1. After the contract of sale has been signed by the customer, the finance manager prepares an FHA firm commitment application form and assembles the following documents:
   a. Three copies of the contract of sale from the finance files
   b. Three copies of complete FHA verification of employment forms from the customer's employer
   c. Three copies of completed FHA bank deposit forms from the customer's bank
   d. Three sets of financial statements from the customer
   e. Three credit references from the customer
   f. Three credit reports on the customer from the credit bureau
2. The FHA firm commitment application is completed by the finance manager, signed, and forwarded with exhibits attached to the mortgagee.
3. The mortgagee approves the application and exhibits and forwards them to the FHA office for acceptance.
4. The application is approved by the FHA office, and a firm commitment is forwarded to the mortgagee.
5. The firm commitment is approved and forwarded by the mortgagee to the finance manager.
6. The finance manager notifies the customer by letter and the marketing manager by copy of the commitment of the FHA approval. The FHA firm commitment form is then forwarded to the office secretary for file in the job folder.

APPROVED BY: _____

DATE: _____

FINANCE PROCEDURE

FROM: General Manager
TO: Finance Manager
SUBJECT: Title Closings

Purpose
The purpose of this procedure is to establish a routine method for handling and processing title closings with customers.

Responsibility
The finance manager is responsible for the coordination of all company activity related to the preparation and processing of title closings with customers.

Title-closing Documents
The following documents must be prepared and available for the title-closing transaction:
 1. Final underwriter's certificate (one copy)
 2. Final survey (one original and two copies)
 3. Insurance policies (one bank and one customer)
 4. Certificate of occupancy (one copy)
 5. Warranties:
    a. Five-year company warranty
    b. One-year roofing warranty
    c. One-year plumbing and heating warranty
    d. Oil burner brochure and warranty
 6. Tax receipts
 7. Closing checks for company attorney
 8. Water company deposit
 9. Home owner's brochure
 10. Customer acceptance inspection report

Procedure
1. The date and time of title closings for each job are established jointly by the construction and finance managers.
2. The finance manager and lender confirm the date and time for the title closing in the lender's office.
3. The finance manager formally advises the company and customer attorneys and the customer in writing of the date, time, and location of the title closing.
4. The insurance agency is advised by copy of the title-closing announcement letter of the closing transaction and is requested to write a policy for the purchaser.
5. A certificate of occupancy is obtained from the town building department by the construction manager.

6. The finance manager checks the title-closing documents for completeness and forwards them to the company attorney for the closing.
7. The title-closing transaction takes place at the office of the lender on the date scheduled and is attended by:
   a. The customer
   b. The customer's attorney
   c. The company attorney
   d. The lender's representatives
   e. A title company representative
8. Following the title-closing transaction, the company attorney forwards a copy of the title-closing statement and escrow agreement (if required) to the finance manager.
9. The title-closing statement and escrow agreement are then forwarded by the finance manager to the office secretary for file in the job folder.

APPROVED BY: _____

DATE: _____

FINANCE PROCEDURE

FROM: General Manager
TO: Finance Manager
SUBJECT: Escrow Agreements

Purpose
The purpose of this administrative procedure is to establish a routine method for the processing of escrow agreements, which are issued when there are delays in the completion of exterior improvements due to the weather or other conditions beyond the control of the company.

Responsibility
The finance manager is responsible for the initiation and processing of escrow agreements with lenders for postponed exterior improvements.

Procedure
1. Prior to the title-closing transaction, the finance manager advises the company attorney that because of conditions beyond the control of the company, exterior improvements will not be completed as scheduled before the closing date.
2. The company attorney advises the lender, and an escrow agreement is prepared for signature at the title closing.
3. The escrow agreement is transacted at the title closing, and a deposit

Finance Procedures

set aside in a special account to ensure completion of, and payment for, the deferred items.
4. Following the title closing, the escrow agreement is forwarded by the company attorney to the finance manager.
5. The escrow agreement is then forwarded by the finance manager to the office secretary for file in the job folder.
6. When the deferred work is completed, the construction manager advises the finance manager and requests that a reinspection be made by the lender.
7. The finance manager orders an inspection of the work completed by the lender.
8. Following the inspection and approval of the deferred work, the lender advises the finance manager that a release of the deposit held in escrow has been authorized.
9. The finance manager submits a certification to the lender for payment of the deposit held in escrow for the deferred work that has been completed.
10. The lender then forwards to the finance manager payment for the deferred work that has been completed.
11. The finance manager records the receipt of the payment with a journal entry, deposits the payment in the bank, and forwards to the office secretary for file in the job folder a memo giving notice that the escrow agreement has been satisfied.

APPROVED BY: _____

DATE: _____

CHAPTER 9

Construction Procedures

CONSTRUCTION PROCEDURE

FROM: General Manager
TO: Construction Manager
SUBJECT: Building Permit Applications

Purpose
The purpose of this procedure is to establish a routine method for the preparation and processing of building permit applications for the authorization of construction operations.

Responsibility
The construction manager is responsible for the preparation and submission of building permit applications with required supporting documents to the town building department.

Documents
The following documents must be submitted to the town building department in order to obtain building permits:
1. Building plans with architect's seal (two sets)

Construction Procedures

2. Building permit application (one copy)
3. Plumbing permit application (one copy)
4. Fireplace permit application (one copy)

Procedure
1. The construction manager is to obtain the necessary application forms for building, plumbing, and fireplace permits from the town building department.
2. The application forms are to be completed by the construction manager and submitted with two sets of architect building plans (with seal attached) to the town building department.
3. The applications are reviewed and approved by the town building department, after which the requested building, plumbing, and fireplace permits are forwarded to the construction manager.
4. The construction manager forwards the permits to the office secretary to make photostatic copies for the project file, after which the original copies are posted in the construction field office.

APPROVED BY: _____

DATE: _____

CONSTRUCTION PROCEDURE

FROM: General Manager
TO: Construction Manager
SUBJECT: Site Preparation

Purpose
The purpose of this procedure is to outline the requirements for the preparation of new sites for land improvement and construction operations.

Responsibility
The construction manager is responsible for the preparation of new sites for land development and construction.

Procedure
1. Before field operations commence at a new construction site, the construction manager is to ensure that the following items are available:
    a. Field office facilities, office equipment, files, forms, etc., required for site administration
    b. Plant personnel for general labor
    c. Job descriptions for field personnel
    d. Procedure manuals for the guidance of field personnel
    e. Site plans and surveys

- f. Construction drawings and specifications
- g. Local building codes and permits
- h. Equipment, instruments, and tools required for site operations
- i. Tool shanties and storage areas for tools and equipment
- j. First-aid kits and safety equipment
- k. Fire extinguishers and emergency equipment
- l. Electricity, water, and telephone equipment
- m. Directional and safety displays
- n. Such other requirements as are deemed necessary for the activation of the particular site

2. The construction manager is to advise the general manager when site preparation requirements have been satisfied.

APPROVED BY: _____

DATE: _____

CONSTRUCTION PROCEDURE

FROM: General Manager
TO: Construction Manager
SUBJECT: Daily Field Reports

Purpose
The purpose of this procedure is to establish a routine method for the preparation and distribution of daily field reports.

Responsibility
The construction manager is responsible for the preparation and submission of daily field reports to the general manager.

Form
Attached is a sample of the daily field report form for use as prescribed in this procedure. The report provides for the transmittal of the following information:
1. Project name
2. Construction manager's name
3. Date of the report
4. Description of the weather
5. Job inspections made
6. Purchase orders issued
7. Subcontractors available
8. Plant labor on hand
9. Equipment rentals
10. Construction manager's signature

Construction Procedures

Preparation
1. Daily field reports are to be prepared in duplicate by the construction manager at the end of each workday and distributed as follows:
    a. One copy (the original) is forwarded to the office secretary for transmittal to the general manager.
    b. One copy (the duplicate) is retained in the construction manager's daily field report file.
2. The office secretary records the receipt of the report in the daily field report register, after which the report is forwarded to the general manager.
3. After the daily field report is reviewed and initialed by the general manager, it is returned to the office secretary for file in the project folder.

APPROVED BY: _____

DATE: _____

DAILY FIELD REPORT

Project_____ Report No._____
Construction Manager_____ Date_____
Weather_____
Subcontractors on Hand_____

Plant Labor_____

Job Inspections_____

Purchase Orders Issued_____

Equipment Rentals_____

Construction Manager_____

CONSTRUCTION PROCEDURE

FROM: General Manager
TO: Office Secretary
SUBJECT: Daily Field Report Register

Purpose
The purpose of the daily field report register is to provide a means for recording and monitoring the receipt of daily field reports issued by the construction manager to the general manager.

Responsibility
The office secretary is responsible for maintaining the daily field report register and documenting therein the receipt of all daily field reports originated by the construction manager.

Register Form
The daily field report register consists of a three-ring binder with loose-leaf register forms for recording the receipt of daily field reports. Attached is a sample register form on which entries are self-explanatory.

Procedure
1. As daily field reports are received by the office secretary, they are chronologically posted in the register.
2. The office secretary is to follow up with the construction manager on the transmittal of daily field reports that have not been received.
3. When the project is completed, the daily field report register is to be inactivated by the office secretary through its physical transfer to the project file.

APPROVED BY: _____

DATE: _____

DAILY FIELD REPORT REGISTER

Date	Report No.	Date	Report No.	Date	Report No.	Date	Report No.

CONSTRUCTION PROCEDURE

FROM: General Manager
TO: Construction Manager
SUBJECT: Job Status Record

Purpose
The purpose of this procedure is to establish a routine method for recording the work status of all jobs under construction and for posting their status on the program control chart.

Responsibility
The construction manager is responsible for the weekly recording of the work status of all jobs under construction.

Record Form
Attached is a completed job status record form used for recording the status of work on all construction jobs. Its preparation, as illustrated, is self-explanatory.

Procedure
1. Each week the construction manager is to personally inspect all jobs under construction and determine the specific work status of each.
2. The construction manager is to post on the job status record, in the blank columns provided, the numbers of each job as they relate to the work activities listed on the form.
3. The work status information posted on the job status record for each job is to be transferred to the program control chart. This is accomplished by plotting in the columns progress bars which relate to the work stage that has been reached as measured by the construction plan drawn on the program control chart.
4. When the construction manager has completed the job status record, it is to be forwarded to the office secretary for file in the project file.

APPROVED BY: _____

DATE: _____

JOB STATUS RECORD

Work Activity Number	Work Activity Description	Number of Workdays	Work Status	
			Job	Job
1-2	Stakeout, clear lot	2.0		
2-3	Excavate foundation	2.0		
3-4	Dig and pour footings	2.0		
4-5	Install foundation walls	4.0		
5-6	Foundation inspection	1.0		
5-7	Parge foundation walls	1.0		
5-9	Install plumbing groundwork	2.0		
5-10	Install water and sewer lines	2.0		
7-8	Tar foundation walls	1.0		
8-11	Backfill foundation	1.0		
9-12	Grade garage floor	1.0		
9-14	Pour cellar slab	2.0		
11-13	Install window wells	1.0		
12-15	Pour garage slab	1.0		
14-16	Set steel girders	1.0		
15-17	Pour garage apron	1.0		
16-18	Install first deck	2.5		
18-19	Clean up debris	1.0		
18-20	Frame first floor and garage	3.5		
20-21	Install second deck	2.0		
21-22	Frame second floor	3.0		
22-23	Install chimney and fireplace	3.0		
22-24	Frame roof	3.0		
24-25	Sheath house	3.0		
25-26	Carpentry inspection	1.5		
25-27	Set exterior door and window frames	1.5		
25-28	Set furnace, tank, water heater	2.0		
25-29	Set stairs	2.0		
25-30	Block out	2.0		
25-31	Rough in heating	2.0		
25-32	Shingle roof	3.0		
25-33	Rough in plumbing	3.0		
27-34	Install exterior doors	1.5		
31-35	Rough in electricity	2.0		
34-39	Install exterior millwork	3.0		
35-36	Connect electric service	1.0		
35-37	Plumbing inspection	1.0		
35-38	Install insulation	2.0		
38-42	Install dry wall	5.0		

JOB STATUS RECORD (Continued)

Work Activity Number	Work Activity Description	Number of Workdays	Work Status Job	Work Status Job
39-40	Apply wood siding	3.0		
40-41	Install garage doors	1.0		
41-43	Dig and pour patio footings	1.0		
41-50	Paint house exterior	4.0		
42-44	Clean up debris	1.0		
42-46	Install plugs and switches	1.5		
42-47	Install heat register and grills	2.0		
42-48	Install ceramic bath tile	2.0		
42-51	Lay hardwood floors	4.0		
43-45	Install patio foundation	1.0		
45-49	Pour patio slab	1.0		
48-53	Install plumbing fixtures	3.0		
50-52	Install gutters and spouts	1.5		
51-55	Install interior doors and trim	4.0		
52-54	Landscape grounds	3.0		
54-58	Install slate walks	1.5		
55-56	Carpentry inspection	1.0		
55-57	Clean up debris	1.0		
55-59	Interior paint	3.0		
58-60	Install stone driveway	2.0		
59-61	Install vanities	2.0		
59-62	Install kitchen appliances	2.0		
59-63	Install foyer and hearth slate	2.0		
59-64	Install kitchen cabinets	2.5		
59-65	Interior decorate	3.0		
64-66	Install mirrors, medicine cabinets	1.0		
64-67	Install shower doors	1.0		
64-68	Install plumbing appliances and trim	1.5		
64-69	Install light fixtures	2.0		
69-70	Electric inspection	1.0		
69-71	Plumbing inspection	1.0		
69-72	Finish hardwood floors	3.5		
72-73	Install resilient floors	2.0		
73-74	Final town inspection	1.0		
73-75	Final bank inspection	1.0		
73-76	Final customer inspection	1.0		
73-77	Customer occupancy permit	1.0		
73-78	Clean house and grounds	4.0		
78-79	Customer acceptance and occupancy	2.0		

CONSTRUCTION PROCEDURE

FROM: General Manager
TO: Construction Manager
SUBJECT: Construction Status Reports

Purpose
The purpose of this procedure is to establish a routine method for monitoring and evaluating performance and progress on construction jobs through the use of weekly construction status reports.

Responsibility
The construction manager is responsible for the preparation and distribution of weekly construction status reports.

Report Form
A sample of the construction status report prescribed in this procedure for monitoring job construction performance is attached. The report form is designed to record the number of workdays variance between actual and scheduled progress dates on the critical path of operations for each job under construction. The number of workdays variance on the critical operations path for each job is posted in the Job Status column of the program control chart in the construction manager's field office. Daily chart postings by the construction manager provide the following up-to-date information for preparing the construction status report:
1. The section and job numbers for all houses now under construction
2. The scheduled start and completion dates for construction performance on each job
3. The actual status of construction performance on each job as measured by the horizontal progress bars posted over the individual job schedules
4. The variance in workdays between the actual and scheduled performance dates reached for each construction job. (The critical path variance entry for the construction status report is posted daily in the Job Status column of the program control chart by the construction manager.)

Procedure
1. Construction status reports are to be prepared in duplicate by the construction manager each Friday from the data posted on the program control chart.
2. The original copy of the report is to be forwarded to the general manager for review with the construction manager on the Monday following its release.
3. The duplicate copy is to be retained by the construction manager in the field office construction status report file for reference and comparison with later reports.

APPROVED BY: _____

DATE: _____

CONSTRUCTION STATUS REPORT

Project   Date

Section No.	Job No.	Job Schedules			Variance +/− Days	Comments
		Start	Status	Finish		

CONSTRUCTION PROCEDURE

FROM: General Manager
TO: Construction Manager
SUBJECT: Construction Schedule Variance Control Reports

Purpose
The purpose of this procedure is to establish a method of monitoring and measuring the variance from schedule for each job under construction.

Responsibility
The construction manager is responsible for maintaining control over the variance from schedule for each job through the use of the construction schedule variance control report.

Report Form
Attached is a copy of the form used for maintaining control over variance from schedule for each construction job. The information required for its preparation is available in the construction status report prepared weekly by the construction manager.

Preparation
1. The variance control report is initiated by the construction manager at the start of the construction program by entering on the report form the project name, its location, the project section number, the number of each construction job, and the scheduled completion dates.
2. Each week the Construction Manager is to post the following information on the report form:
   a. The posting date at the top of the vertical column for the variance to be entered
   b. Beneath the posting date entry, the number of workdays variance ahead (+) or behind (−) is noted for each job under construction
3. The weekly variance postings on the report form will provide a variance trend for each construction job. It is to be used by the construction manager for controlling construction progress on each job and for monitoring overall construction performance in the program.

APPROVED BY: _____

DATE: _____

CONSTRUCTION SCHEDULE VARIANCE CONTROL											
Section No.	Job No.	Schedule Complete	Weekly Job Variance from Schedule								
			/	/	/	/	/	/	/	/

CONSTRUCTION PROCEDURE

FROM: General Manager
TO: Construction Manager
SUBJECT: Customer Complaints

Purpose
The purpose of this procedure is to establish a routine method for processing and expediting the correction of customer complaints.

Responsibility
The construction manager is responsible for processing and recording all customer complaints.

Complaint Form
Customer complaints may originate either before or after the completion of the job and the closing of title to the house. Whether the complaint is received in person from the customer, or by correspondence or telephone call, the customer must complete the complaint form. A copy of the form is attached and its completion requires the following information:
 1. Customer's name
 2. Address of customer
 3. Telephone number of customer
 4. Date of customer complaint
 5. Job number—customer house number
 6. Project—where customer's house is located
 7. Location—if a custom house off the project site
 8. Description of complaint—a customer write-up on the nature of the complaint
 9. Customer's signature
10. Action taken by construction manager
11. Construction manager's signature
12. Date of construction manager's signature

Procedure
1. When the customer presents the construction manager with a complaint concerning the construction of his house, the construction manager requests the customer to complete a customer complaint form in the field office.
2. The complaint form is completed by the customer in triplicate and reviewed with the construction manager.
3. The construction manager then inspects the job complaint with the customer.
4. If the complaint is not justified, the construction manager advises the customer, notes the reason on the complaint forms, and distributes them as follows:

     a. One copy to the office secretary for file in the job folder
     b. One copy to the customer
     c. One copy for the field office customer complaint file
5. If the complaint is justified, the construction manager indicates the reason on the complaint forms and distributes them as follows:
     a. One copy to the customer
     b. One copy to the field office customer complaint follow-up file
     c. One copy to be used for preparing a job work order
6. The construction manager issues a job work order to remedy the complaint (see job work order procedure).
7. After the complaint has been corrected, the action taken is noted on the job work order copy of the complaint form which is then forwarded to the office secretary for file in the job folder. The follow-up copy of the complaint form is then filed in the field office customer complaint file.

APPROVED BY: _____

DATE: _____

CUSTOMER COMPLAINT

Customer_____ Date_____ Job No._____
Address_____ Project_____
Phone No._____ Location_____

Description of Complaint

_____
_____
_____
_____
_____
_____
_____
_____
_____
_____
_____

                                                                  Customer_____

Action Taken

_____
_____

                                      Construction Manager_____
                                                                Date_____

CONSTRUCTION PROCEDURE

FROM: General Manager
TO: Construction Manager
SUBJECT: Customer Complaint Control

Purpose
The purpose of this procedure is to establish an effective means for monitoring and controlling the processing of customer complaints.

Responsibility
The construction manager is responsible for establishing and maintaining control over all customer complaints received and processed.

Control Form
Attached is a sample of the customer complaint control form to be used as prescribed in this procedure. The form makes provision for posting the following information:
1. Date received—the date that the complaint is received from the customer
2. Date completed—the date that the complaint is resolved
3. Job number—the number of the customer's construction job
4. Complaint—a brief description of the complaint
5. Work order number—the number of the work order issued to correct the complaint
6. Work trade—the subcontractor involved
7. Action taken—a brief description of action taken

Procedure
1. The customer complaint control form is to be attached to the inside cover of the field office customer complaint folder.
2. When complaint forms are completed by the customer, they are immediately recorded on the customer complaint control form by the construction manager.
3. If the construction manager determines after an inspection of the customer's complaint that it is not justified, the findings are recorded on the customer complaint control form.
4. If a work order is issued to correct the complaint, the construction manager makes a record of the action taken on the customer complaint form after the complaint is remedied.

APPROVED BY: _____

DATE: _____

CUSTOMER COMPLAINT CONTROL

Date Received	Action Date	Job No.	Complaint	Work Order No.	Work Trade	Action Taken

CONSTRUCTION PROCEDURE

FROM: General Manager
TO: Construction Manager
SUBJECT: Job Work Orders

Purpose
The purpose of this procedure is to establish a work order system through which subcontractors may be issued work directives to remedy defective workmanship or to perform additional construction work required by changes in building plans and specifications.

Responsibility
The construction manager is responsible for the issuance of all job work orders and the subsequent follow-up action required to ensure that work deficiencies are remedied or changes in work requirements are carried out as directed.

Work Order Form
Attached is a sample of the job work order form to be used as prescribed in this procedure. For completion the form requires the following information:
 1. Work order number—work orders are issued in numerical sequence
 2. Date—on which the work order was issued
 3. Job number—that the work order was issued for
 4. Project—where the job is located
 5. Location—of jobs located off the project site
 6. Subcontractor—who is to receive the work order
 7. Address—of the subcontractor
 8. Work description—the work to be performed by the subcontractor
 9. Subcontractor's comments—may apply to extra charges or problems encountered by the subcontractor
 10. Work completed:
    a. Subcontractor's signature and date work completed
    b. Owner's signature and date work accepted (for rework required as a result of deficiencies uncovered during customer acceptance inspections, or after customer has taken title to the house)
    c. Construction manager's signature and date work accepted

The job work order form must be fully completed to ensure that work is performed as required and to avoid unnecessary subcontractor call-backs.

Procedure
1. Job work orders may result from any one of the following actions:
   a. FHA compliance inspections
   b. Lender compliance inspections
   c. Town building compliance inspections

    d. Town health compliance inspections
    e. Customer acceptance inspections
    f. Company quality control inspections
    g. Changes in construction plans and specifications
2. Job work orders are to be prepared in triplicate by the construction manager and distributed as follows:
    a. Original and one copy are forwarded to the subcontractor.
    b. One copy is filed in a job work order follow-up folder in the field office.
3. The construction manager follows up on the subcontractor's compliance with work orders by daily job inspections.
4. The subcontractor carries out the work assignment as specified, signs the work order, and notes thereon the date of its completion.
    a. If the work order was issued to correct a customer complaint, the customer must initial and date the work order to indicate acceptance.
    b. If the work order is to correct defective workmanship during construction, it is to be returned to the construction manager after the work is completed. The work will then be inspected by the construction manager, and the work order signed and dated to indicate acceptance.
5. After the work completed is approved, the construction manager returns the original copy of the work order to the subcontractor and retains the extra copy.
6. If a subcontractor charge is incurred as a result of the work performed, the copy endorsed by the construction manager is forwarded to finance for payment. The subcontractor charges are noted on the follow-up copy of the work order by the construction manager, and it is forwarded to the office secretary for file in the job folder.
7. If subcontractor charges are not incurred against the work order, the signed copy is forwarded by the construction manager to the office secretary for file in the job folder, and the field follow-up copy is then destroyed.

APPROVED BY: _____

DATE: _____

JOB WORK ORDER

Date_____

Project_____ Job No._____
Location_____ Work Order No._____
Subcontractor_____
Address_____

Work Description

_____
_____
_____
_____
_____
_____
_____
_____
_____

Subcontractor's Comments

_____
_____

WORK COMPLETED:　Subcontractor_____　Date_____
　　　　　　　　　Owner_____　Date_____
　　　　　　　　　Construction Manager_____　Date_____

CONSTRUCTION PROCEDURE

FROM: General Manager
TO: Construction Manager
SUBJECT: Certificate of Occupancy Applications

Purpose
The purpose of this procedure is to establish a routine method for the preparation and processing of certificate of occupancy applications.

Responsibility
The construction manager is responsible for the coordination of company activities and the assembling of documents required to make applications for certificates of occupancy with local building authorities.

Documents
The following documents must be secured for submittal with applications for certificates of occupancy:
1. Final health department approval (two copies)
2. Final survey (three copies)
3. Final underwriter's certificate (one copy)
4. Final town inspection approval reports for carpentry, plumbing, and such other inspections as may be required

Procedure
1. The construction manager is to obtain the application form for the certificate of occupancy from the town building department.
2. Final health department approvals, final surveys, final underwriter's certificate, and final town inspection approvals are secured by the construction manager from their respective issuing sources.
3. The certificate of occupancy application form is completed by the construction manager and forwarded with copies of the final survey, final health department approval, final underwriter's certificate, and final town inspection report to the town building department.
4. After approval of the application, the town building department forwards the approved certificate of occupancy to the construction manager.
5. The certificate of occupancy is forwarded by the construction manager to the office secretary for reproduction and distribution as follows:
   a. One photostatic copy is retained in the project file.
   b. The original copy is forwarded to the finance manager for use in the closing of title with the customer.

APPROVED BY: _____

DATE: _____

CHAPTER 10

Purchasing Procedures

PURCHASING PROCEDURE

FROM: General Manager
TO: Construction Manager
SUBJECT: Requests for Quotation

Purpose
The purpose of this procedure is to establish a routine method for the preparation and processing of requests for quotation issued to subcontractors and suppliers.

Responsibility
The construction manager is responsible for the preparation, issuance, and processing of all requests for quotation.

RFQ Form
Attached is a sample of the request for quotation (RFQ) form prescribed for use in this procedure. The RFQ form must be fully completed by subcontractors and suppliers to be accepted. The construction manager is to maintain a qualified-bidders' file of subcontractors and suppliers who best meet the company's quality, pricing, and delivery requirements.

Procedure

1. As directed by the construction manager, the office secretary is to prepare request for quotation forms in triplicate for selected subcontractors and suppliers.
2. After completing the request for quotation forms, the office secretary submits them to the construction manager for approval and signature.
3. The construction manager approves and signs the request for quotation forms, attaches specifications and drawings (if required), and returns them to the office secretary.
4. The original and one copy of the request for quotation are forwarded to subcontractors and suppliers by the office secretary, and one copy is retained in the RFQ follow-up file.
5. Subcontractors and suppliers are to inspect the model house for quality of workmanship and materials, study the specifications and drawings, then complete the RFQ forms and return them to the office secretary.
6. The returned RFQ forms are then forwarded by the office secretary to the construction manager for review and source selection.
7. After the successful bidders are selected, their requests for quotation and bids are returned to the office secretary for use in preparing purchase orders or subcontractor agreement drafts.
8. The office secretary advises unsuccessful bidders by letter that awards have been placed with other sources.
9. RFQ forms for successful bidders are filed by the office secretary with purchase order copies in the project file. Those for unsuccessful bidders are filed in the project file RFQ folders.

APPROVED BY: _____

DATE: _____

REQUEST FOR QUOTATION	THIS IS NOT AN ORDER
From_____ _____ _____ To_____ _____ _____	Bids Close_____ Request No._____ Date_____ 19__ Date Requested_____ F.O.B._____ Job No._____

Location of Model

Please quote your best price and delivery on the items listed below:
To be shipped via_____ To_____
Terms_____ Discount_____ Days, Net_____ Days____

Quotations to be submitted in accordance with the terms and conditions on the reverse side of this form.

Quantity	Description (substitutes must be clearly shown)	Price

Please return carbon copy promptly with information completed. Original copy is for your records.

By_____

PURCHASING PROCEDURE

FROM: General Manager
TO: Construction Manager
SUBJECT: Purchase Orders

Purpose
The purpose of this procedure is to establish a routine method for the preparation and processing of purchase orders.

Responsibility
The construction manager is responsible for the preparation, issuance, and processing of all purchase orders for construction purposes.

Purchase Order Form
Attached is a sample of the purchase order form prescribed for use in this procedure. Purchase orders must be prepared and issued for all construction procurements.

Procedure
1. Purchase orders are prepared in quadruplicate by the construction manager and distributed as follows:
   a. The original and acknowledgment copies are forwarded to the supplier.
   b. One copy is retained by the construction manager to match with shipping documents accompanying incoming materials.
   c. One copy is forwarded to the office secretary for follow-up action with the supplier.
2. When the materials ordered have been received from the supplier at the construction site, the construction manager matches the field copy of the purchase order against the shipping documents received with the delivery.
3. The construction manager records receipt and approval of the materials ordered by initialing the field office copy of the purchase order, which is then forwarded with the shipping documents to the office secretary.
4. The office secretary files the follow-up copy of the purchase order in the project file. The field copy of the purchase order and shipping documents are then forwarded to the finance manager for use in auditing incoming supplier invoices.

APPROVED BY: _____

DATE: _____

From_____ PURCHASE ORDER _____	This Order Number Must Appear on All Invoices, Packing Lists, Shipments, and Correspondence.
To_____ _____ _____ Ship To_____ _____	P.O. No._____ Date_____ 19____ Ship Via_____ Date Requested_____ Job No._____ Terms_____

Quantity	Description	Price	Total

For_____ By_____

This contract subject to conditions printed on the reverse of this order.

PURCHASING PROCEDURE

FROM: General Manager
TO: Office Secretary
SUBJECT: Purchase Order Register

Purpose

The purpose of this procedure is to establish a routine method for recording and controlling the issuance of purchase orders by means of a purchase order register.

Responsibility

The office secretary is responsible for maintaining the purchase order register and recording therein the preparation and release of all purchase orders.

Register Form

The purchase order register consists of a three-ring binder with loose-leaf register forms for recording data on all purchase orders released. Attached is a sample register form on which provision is made for recording the date, number, supply source, description, name of originator of purchase order, and dates that supplies are required and received.

Procedure

1. Upon receipt of the office copy of the purchase order, the office secretary records the date, number, vendor, description, and name of originator in the register.
2. The office secretary makes use of the register to follow up on delinquent suppliers for delivery of construction materials and supplies.
3. Upon receipt of the construction manager's field office copy of the purchase order with supplier shipping documents, the office secretary records the date of delivery of the supplies in the purchase order register.
4. When the project is completed, the purchase order register is inactivated by the office secretary's transfer of the register to the project file.

APPROVED BY: _____

DATE: _____

PURCHASE ORDER REGISTER

Purchase Order No.	Date	Vendor	Description	Originator	Dates	
					Requested	Received

PURCHASING PROCEDURE

FROM: General Manager
TO: Construction Manager
SUBJECT: Vendor Certifications

Purpose

The purpose of this procedure is to establish a routine method for certifying that supplies and materials have been received from vendors and are as specified in purchase orders before the invoices are paid.

Responsibility

The construction manager is responsible for checking vendor shipping documents against purchase orders to ensure that deliveries made are in accordance with quality and quantity requirements specified.

Procedure

1. Vendor invoices for supplies and materials are to be paid by the finance department on the tenth working day of each month. All vendor invoices are to be forwarded to the construction manager for approval.
2. Before approving invoices, the construction manager checks the purchase order register to ascertain that vendor shipments have been received, after which the file copy of the purchase order in the project file is reviewed to ensure that no exceptions were noted for the shipment.
3. If vendor shipments have been completed and no exceptions taken, the construction manager is to note approval for payment by initialing and dating the invoice and returning it to the finance department.
4. If exceptions have been taken on the quality and/or quantity of the vendor's shipments. the invoice is not to be paid until the exception has been remedied by the vendor.

APPROVED BY: _____

DATE: _____

PURCHASING PROCEDURE

FROM: General Manager
TO: Office Secretary
SUBJECT: Vendor Register

Purpose
The purpose of this procedure is to establish a vendor register for listing the names, addresses, and telephone numbers of vendors selected for furnishing construction material and supply requirements on the project.

Responsibility
The office secretary is responsible for the preparation and maintenance of the vendor register.

Register Form
The vendor register consists of a three-ring binder with loose-leaf register forms for recording the names, resources, addresses, and telephone numbers of vendors.

Procedure
1. Upon receipts of purchase orders for vendors selected for furnishing construction material requirements, the office secretary is to record in the vendor register the name, address, and telephone numbers of all vendors.
2. Those vendors that submitted unsuccessful bid quotations are to be listed by supply category on separate register pages:

   a. Concrete
   b. Lumber
   c. Millwork
   d. Kitchen cabinets
   e. Appliances
   f. Steel
   g. Exterior hardware
   h. Interior hardware
   i. Kitchen accessories
   j. Bath accessories
   k. Miscellaneous

3. Inactive vendor lists are to be used for the selection of additional or replacement vendors as may be needed to meet construction requirements.
4. When the project is completed, the vendor register is to be inactivated by its transfer to the project file.

APPROVED BY: _____

DATE: _____

VENDOR REGISTER

Name	Materials	Address	Telephone

CHAPTER 11

Subcontracting Procedures

SUBCONTRACTING PROCEDURE

FROM: General Manager
TO: Construction Manager
SUBJECT: Subcontractor Agreements

Purpose
The purpose of this procedure is to establish a routine method for the negotiation and processing of subcontractor agreements.

Responsibility
The construction manager is responsible for the negotiation, preparation, distribution, and administration of all subcontractor agreements. The general manager is responsible for approving and signing all subcontractor agreements.

Agreement Form
Subcontractor agreements are to be documented and executed on the company's standard subcontractor agreement form.

Procedure
1. All subcontractor agreements are to be definitized before the end of the second week following the start of construction operations.
2. The construction manager is to forward rough drafts of subcontractor agreements with negotiated prices, payments, performance terms, and conditions to the office secretary for the preparation of originals with three copies.
3. The office secretary is to forward the final subcontractor agreement forms to the general manager for review, approval, and return.
4. Two copies of the approved subcontractor agreements are forwarded by the office secretary to the subcontractor for approval, signature, and the return of both with three copies of certificates of insurance.
5. If no exceptions have been noted by the subcontractor, the subcontractor agreements are executed by the general manager and distributed by the office secretary as follows:
   a. One copy (the original) is filed in the project folder.
   b. One copy marked "executed" is forwarded to the finance manager.
   c. One copy marked "executed" is forwarded to the construction manager.
   d. One copy (executed) is returned to the subcontractor.
6. Certificate of insurance copies are distributed by the office secretary as follows:
   a. One copy is filed in the project folder with the subcontractor agreement.
   b. One copy is forwarded to the finance manager.
   c. One copy is forwarded to the company's insurance broker.
7. If exceptions are noted on the agreements by the subcontractor, they are reviewed by the general manager and if approved, returned to the subcontractor for execution as amended. After it has been executed by the subcontractor, the amended agreement is distributed as outlined in the preceding paragraphs of this procedure.
8. Subcontractor performance is to be monitored by the construction manager in conformance with the terms and conditions specified in the subcontractor agreement.

APPROVED BY: _____

DATE: _____

SUBCONTRACTING PROCEDURE

FROM: General Manager
TO: Construction Manager
SUBJECT: Subcontractor Work Schedules

Purpose
The purpose of this procedure is to establish an effective system for the preparation, issuance, and control of subcontractor work schedules.

Responsibility
The construction manager is responsible for establishing and controlling subcontractor work schedules for each job under construction.

Schedule Form
Attached is a sample of the subcontractor work schedule form to be used as prescribed in this procedure. The form documents the following information:

1. Date—that the schedule is prepared and issued
2. Project—where the job is located
3. Location—of jobs located off the project site
4. Subcontractor—for whom the schedule is prepared
5. Address—of the subcontractor
6. Work description—that particular activity performed by the subcontractor
7. Job number—the job on which the subcontractor will work
8. Schedule dates:
    a. Start—the dates that the subcontractor is scheduled to begin work on each job
    b. Finish—the dates that the subcontractor is scheduled to complete work on each job
9. Variance +/− workdays:
    a. Start—the number of construction workdays ahead or behind schedule that the subcontractor starts work on each job
    b. Finish—the number of construction workdays ahead or behind schedule that the subcontractor completes work on each job
10. Comments—special requirements associated with the work to be performed on each job

The program control chart located in the construction manager's field office provides the necessary schedule information required to complete the subcontractor work schedule form.

Procedure
1. The construction manager is to prepare the subcontractor work schedule in duplicate. The subcontractor's name, the list of jobs on which work

is to be performed, and the scheduled start and completion dates for their work activities as shown on the program control chart are posted on the subcontractor's work schedule form.
2. When a subcontractor's work activities on a particular job are not performed concurrently, or not performed in sequence one right after the other, then separate schedule forms must be completed for each work activity. To illustrate, the paint subcontractor may have the contract to perform both interior and exterior painting. If both these work activities are not performed at the same time (concurrently and in parallel with each other), or one right after the other (sequentially), then separate schedules must be prepared for each work activity. One schedule form must be prepared for interior painting, and another for exterior painting.
3. After the schedule dates for work to be performed are entered on the schedule form, any special instructions that may apply to particular jobs must be noted by the construction manager in the Comments column provided.
4. The completed schedule form is then distributed as follows:
   a. The original copy is given to the subcontractor by the construction manager, and the subcontractor instructed in its use.
   b. The duplicate copy is retained by the construction manager in the field office subcontractor work schedule file.
5. After the subcontractor work schedule has been issued, the construction manger is to record weekly in the Variance columns of the schedule form the number of days that the subcontractor is ahead or behind schedule as of that date. The program control chart shows the cumulative gain or slippage of the subcontractor's performance as of the latest date of entry on the chart, while the subcontractor work schedule presents a historical record of the subcontractor's performance on the project.
6. Subcontractor work schedules are to be transferred from the field office file to the project file by the construction manager when subcontractors have completed their scheduled assignments.

APPROVED BY: _____

DATE: _____

SUBCONTRACTOR WORK SCHEDULE

Project_____

Location_____

Date_____

Subcontractor_____
Address_____
Work Description_____

Work Schedule

Job No.	Schedule Dates		Variance +/− Days		Comments
	Start	Finish	Start	Finish	

SUBCONTRACTING PROCEDURE

FROM: General Manager
TO: Construction Manager
SUBJECT: Subcontractor Certifications

Purpose
The purpose of this procedure is to establish a routine method for certifying that subcontractors have performed in accordance with the terms and conditions of their subcontractor agreements before payment is remitted for their services.

Responsibility
The construction manager is responsible for inspecting work completed by subcontractors, certifying that services have been rendered as specified in agreements, and approving subcontractor invoices for payment.

Procedure
1. Subcontractor invoices for services rendered are to be paid by the finance department on the tenth working day of each month. All invoices are to be forwarded to the construction manager for approval.
2. Before approving invoices, the construction manager is to inspect the quality of the subcontractor's work to certify that it meets the company's construction standards and the terms and conditions stipulated in the subcontractor's agreement.
3. After inspecting the work for which the invoice was submitted, the construction manager is to note approval for payment by initialing and dating the invoice and returning it to the finance department.
4. If the construction manager takes exception to the quality of the subcontractor's workmanship, the invoice is not to be paid until the exception has been remedied by the subcontractor.

APPROVED BY: _____

DATE: _____

SUBCONTRACTING PROCEDURE

FROM: General Manager
TO: Office Secretary
SUBJECT: Subcontractor Register

Purpose
The purpose of this procedure is to establish a subcontractor register for listing the names, addresses, and telephone numbers of subcontractors engaged in construction activities on the project.

Responsibility
The office secretary is responsible for the preparation and maintenance of the subcontractor register.

Register Form
The subcontractor register consists of a three-ring binder with loose-leaf register forms for recording the names, trades, addresses, and telephone numbers of subcontractors.

Procedure
1. Upon receipt of subcontractor agreements for the file, the office secretary is to record in the subcontractor register the name, trade, address, and telephone number of all subcontractors.
2. Those subcontractors that submitted unsuccessful bid quotations are to be listed on separate register forms in the following trade categories:

   a. Masonry
   b. Rough carpentry
   c. Plumbing-Heating
   d. Excavation
   e. Insulation
   f. Painting
   g. Ceramic tile
   h. Landscaping
   i. Roofing
   j. Siding
   k. Electrical
   l. Surveyors
   m. Dry wall
   n. Hardwood floors
   o. Finish carpentry
   p. Resilient tile

3. Inactive subcontractor lists are to be used for the selection of additional or replacement subcontractors as may be needed to meet construction requirements.
4. When the project is completed, the subcontractor register is to be inactivated by its transfer to the project file.

APPROVED BY: _____

DATE: _____

SUBCONTRACTOR REGISTER

Name	Trade	Address	Telephone

CHAPTER 12

Inspection Procedures

INSPECTION PROCEDURE

FROM: General Manager
TO: Construction Manager
SUBJECT: Town Building Department Inspections

Purpose
The purpose of this procedure is to establish a routine method for handling town building department inspections.

Responsibility
The construction manager is responsible for the scheduling and coordination of all town building department inspections required for each job during the course of construction.

Inspection Forms
The inspection forms furnished by the town building department are to be used for scheduling and requesting inspections as follows:

1. First inspection—foundation in place
2. First carpentry inspection—house framed and sheathed

3. First plumbing inspection—rough plumbing completed
4. Final carpentry inspection—interior trim installed
5. Final plumbing inspection—plumbing trim completed
6. Final house inspection—house completed

The town building department inspection forms must be obtained at the time the construction permit is received for each job and must be held by the office secretary for release for inspections when requested by the construction manager.

Procedure
1. The construction manager schedules town building department inspections by furnishing the office secretary with a list of jobs to be inspected, indicating specific inspections required and the dates for each.
2. The office secretary completes the inspection request forms and forwards them to the town building department.
3. The town building department inspector conducts the inspections requested, completes the inspection request forms indicating approval or rejection, endorses the forms, and returns them to the office secretary.
4. The inspection forms are then forwarded by the office secretary to the construction manager for information and recording on the construction inspection record.
5. The construction manager returns the inspection forms to the office secretary to file in the job folder.

APPROVED BY: _____

DATE: _____

INSPECTION PROCEDURE

FROM: General Manager
TO: Construction Manager
SUBJECT: Town Health Department Inspections

Purpose
The purpose of this procedure is to establish a routine method for obtaining water lateral, well, and cesspool inspections as required by the town health department.

Responsibility
The construction manager is responsible for ensuring that subcontractors obtain the inspections and subsequent approvals for all water lateral, well, and cesspool installations in accordance with town health department regulations.

Inspection Procedures

Inspection Forms

The forms required for town health department inspections and approvals are obtained and processed by the subcontractors responsible for the installation of water laterals, wells, and cesspools.

Procedure

A. *Water lateral inspections*
   1. After water laterals are installed, the construction manager instructs the office secretary to request the plumbing subcontractor to secure town health department approval.
   2. The office secretary requests the plumber to secure town health department approval for the water lateral installation.
   3. The plumber requests town health department inspection and approval of water laterals.
   4. Town health department inspects water lateral installation and forwards to the plumber its letter of approval.
   5. The plumber advises the office secretary that water laterals have been approved by the town health department.
   6. The office secretary advises the construction manager that the water laterals have been approved by the town health department.
   7. Backfill of the water lateral trench is ordered by the construction manager.
   8. When the water meters are installed and the water is tapped, the town water authority informs the construction manager by letter (with two extra copies).
   9. The construction manager forwards the copies of the letter to the health department requesting its approval of the water installation and sends the original copy of the letter to the office secretary for file in the job folder.
   10. After making inspection, the health department sends the construction manager a letter of approval for the water installation.
   11. The construction manager records the approval in the construction inspection record and forwards the letter to the office secretary for file in the job folder.

B. *Well inspections*
   1. After the water well has been installed, the construction manager instructs the office secretary to request the well driller subcontractor to secure the preliminary approval of the town health department.
   2. The office secretary requests the well driller to obtain health department approval.
   3. The well driller requests health department inspection and preliminary approval of the well.
   4. The health department inspects the well and grants the well driller preliminary approval.
   5. The well driller advises the office secretary that preliminary health department approval of the well has been obtained.
   6. The office secretary advises the construction manager that preliminary health department approval of the well has been received.

7. The construction manager authorizes the well driller to install the pump and electricity in the well.
8. The well driller advises the construction manager when the pump and electricity are installed in the well.
9. The construction manager obtains three copies of a laboratory analysis of the well water and instructs the office secretary to file final well approval application with the health department.
10. The office secretary submits the application for the final well approval to the health department with:
    a. One copy of the water analysis report
    b. One copy of a well driller's affidavit
    c. The well number (furnished by well driller)
11. The health department inspects the well and forwards its letter of approval to the construction manager.
12. The construction manager forwards to the office secretary:
    a. One copy of the water analysis report
    b. One copy of the well driller's affidavit
    c. The health department's letter of approval for the well
    The office secretary files the documents in the job folder.

C. *Cesspool inspections*
1. When the building permit is received, the office secretary furnishes the cesspool subcontractor information on:
    a. The job number
    b. The size of the lot
    c. The location of the lot
    d. The number of the building permit
2. After the cesspool is installed, the subcontractor applies to the health department for approval.
3. The health department inspects the cesspool and plot plan to be sure that the cesspool meets specifications and is not improperly located under walks, then grants approval.
4. The subcontractor advises the office secretary that the cesspool has been approved by the health department.
5. The office secretary informs the construction manager of the approval.
6. The construction manager authorizes the cesspool subcontractor to backfill excavation.

APPROVED BY: _____

DATE: _____

INSPECTION PROCEDURE

FROM: General Manager
TO: Construction Manager
SUBJECT: FHA Compliance Inspections

Purpose
The purpose of this procedure is to establish a uniform method for processing all FHA compliance inspections.

Responsibility
The construction manager is responsible for the scheduling and coordination of all FHA compliance inspections.

Inspection Request Form
The office secretary maintains standard FHA request for compliance inspection forms, which are furnished by the mortgagee and issued as directed by the construction manager.

Procedure
A. *First FHA compliance inspection*
   1. When foundations are completed and ready for backfill, the construction manager forwards to the office secretary a list of jobs to be inspected with their FHA serial numbers and the dates that they will be ready for inspection.
   2. The office secretary completes request for compliance inspection forms furnished by the mortgagee and forwards them to the local FHA office.
   3. The first FHA compliance inspection is conducted by the FHA field inspector with the construction manager on the date scheduled.
   4. The FHA field inspector furnishes the construction manager with a field copy of the inspection report and submits official copies to the FHA office.
   5. The construction manager records the job inspections made in the construction inspection record and forwards the field copy of the FHA compliance inspection report to the finance manager.
   6. The finance manager makes duplicates of the field copy of the inspection report and forwards them to the mortgagee with a letter of transmittal certifying that work has been completed and requesting payment.
   7. The FHA office forwards an approval copy of the official compliance inspection report to the mortgagee.
   8. The mortgagee forwards the official compliance inspection report to the finance manager.

9. The official compliance inspection report is routed by the finance manager to the office secretary for file in the job folder.

B. *Second FHA compliance inspection*

When houses are enclosed and roofed, structural framing completed and exposed, and roughing in of heating, plumbing, and electric work is completed and visible for inspection, the construction manager forwards to the office secretary a list of those jobs to be inspected with their FHA serial numbers and the dates that they will be ready for inspection.

(*Note:* The procedure followed for this second FHA compliance inspection is the same as that for the first inspection.)

C. *Third FHA compliance inspection*

When construction improvements are thoroughly completed and houses are ready for occupancy, the construction manager forwards to the office secretary a list of those jobs to be inspected with their FHA serial numbers and the dates that they will be ready for inspection.

(*Note:* The procedural steps for the third FHA compliance inspection are the same as those followed for the first and second inspections with the exception that in step 8 the mortgagee forwards a letter of acceptance with the official compliance inspection report to the finance manager.)

APPROVED BY: _____

DATE: _____

INSPECTION PROCEDURE

FROM: General Manager

TO: Construction Manager

SUBJECT: Lender Compliance Inspections

Purpose

The purpose of this procedure is to establish a schedule for lender inspections of work completed in order to obtain payments on construction loans as agreed upon between the company and the lender.

Responsibility

The construction manager is responsible for ensuring that the company has complied with the construction plans, specifications, and minimum standards as required, and for scheduling and ordering compliance inspections by the lender as predesignated work stages are completed.

Schedule of Lender Inspections
1. *First inspection:* rough enclosure—30 percent of building loan. The building should be framed and sheathed, the finished roof laid and completed, and window and exterior door frames set. Underflooring should be laid, and partition work should be set. Guaranteed survey submitted showing outside lines and projections to roof (foundation survey).
2. *Second inspection:* full enclosure—30 percent of building loan. Exterior finish of brick veneer, siding, shingles, or stucco must be completed and all exterior woodwork primed. Chimney must be built, all rough plumbing and electric BX wiring installed, and stairs up. Lathing should be completed, window sash should be installed, and brown mortar or plaster completed. If sheetrock is used, it must be in place and all nail holes and joints spackled or plastered. All plastering must be completed. Water and sewer connection to street must be completed, or cesspool connected. Bathtub should be set. Concrete cellar floor should be laid.
3. *Third inspection:* fixtures—20 percent of building loan. Tile work must be completed, and plumbing fixtures installed. Gutters and leaders must be installed. Heating boiler must be connected, domestic hot-water heating equipment installed, and kitchen cabinets hung. Finish floor should be laid, insulation installed, and all trim completed.
4. *Fourth inspection:* final payment—20 percent of building loan. The exterior woodwork should have the final coat of paint. Decorating must be completed, linoleum laid, and floors finished. Heating burner, radiators, range hardware, and lighting fixtures must be completely installed. Sidewalks, steps, and terraces must be completed. Grading should be completed, topsoil applied, and shrubbery and lawn planted. Curbs should be in and street paved. The building should be entirely completed as called for in plans and specifications and ready for occupancy. All kitchen appliances should be installed. Gas and electric service connections must be made.

Inspection Forms
The forms required for lender compliance inspections are held by the lender, and copies are given to the construction manager in the field after each inspection is completed.

Procedure
1. Before the building is rough enclosed, the construction manager telephones the lender and furnishes the date that the first stage of construction will be completed and ready for compliance inspection.
2. The first-stage compliance inspection is made by the lender and the construction manager on the date scheduled.
3. The lender furnishes the construction manager with two copies of the compliance inspection report and authorizes construction to proceed.
4. The construction manager records compliance inspections in the construction inspection record and forwards the completed inspection reports to the finance manager.

5. The finance manager certifies to the lender that work has been completed and requests payment. One copy of the inspection report is forwarded to the office secretary for file in the job folder.

(*Note:* The same inspection procedure is followed for lender compliance inspections at the completion of the second, third, and fourth construction stages.)

APPROVED BY: _____

DATE: _____

INSPECTION PROCEDURE

FROM: General Manager
TO: Construction Manager
SUBJECT: Customer Acceptance Inspections

Purpose
The purpose of this procedure is to establish a routine method for conducting customer acceptance inspections before and after the title closing.

Responsibility
The construction manager is responsible for scheduling and conducting acceptance inspections with customers and for the subsequent preparation of customer acceptance inspection reports.

Report Form
Attached is a copy of the customer acceptance inspection report to be completed by the construction manager. The following information must be entered on the form:

1. Project—where the job is located
2. Job—number of the customer's house
3. Customer's name.
4. Phone—where customer may be contacted
5. Date—that the inspection was made
6. Inspection list—items to be checked with customer
7. Approval—check *yes* or *no* for customer approval
8. Comments—items requiring rework
9. General remarks—cover items not on the inspection list
10. Customer—signature of customer and spouse
11. Construction manager's signature

Inspection Procedures 181

Procedure
A. *Closing inspections*
1. The construction manager directs the office secretary to set up an appointment for the customer to make the closing inspection.
2. The customer is informed by the office secretary of the time and date for the closing inspection.
3. The customer attends the closing inspection with the construction manager, and they jointly inspect the house and the lot.
4. The construction manager completes the customer acceptance inspection report, checking each inspection item for customer approval or disapproval and noting comments pertinent to the rework of items inspected. After the customer (and spouse) sign the report, it is endorsed by the construction manager and distributed as follows:
   a. One copy is furnished the customer.
   b. One office copy is for the office secretary's job folder.
   c. One field copy is used to issue job work orders.
   The inspection is then entered on the construction inspection record.
5. The construction manager prepares job work orders (see work order procedure) for subcontractors to repair defective work and files the field copy of the customer acceptance inspection report in a follow-up folder.
6. The subcontractors repair defective work, obtain the customer's signature on the job work orders, and return the orders to the construction manager.
7. When all the customer complaints as noted on the customer acceptance inspection report have been corrected, the dates that they were made are recorded on the report, and it is then forwarded to the office secretary for file in the job folder.
8. A duplicate copy of the report with the corrections noted thereon is forwarded to the company attorney for the title closing.
   If rework is not completed before the closing, the file copy of the report is forwarded to the attorney, and an escrow agreement is drawn to cover work not completed.
B. *Postclosing inspections*
1. The construction manager conducts a postclosing inspection of the house and lot with the customer one month after the closing.
2. A new customer acceptance inspection report is prepared, and the aforementioned procedural steps (1 to 7) for closing inspections are repeated.

APPROVED BY: _____

DATE: _____

CUSTOMER ACCEPTANCE INSPECTION REPORT

Project_____ Job No._____
Customer_____ Phone_____ Date_____

Inspection List		Approve Yes	Approve No	Comments
1. Doors	Finish			
	Work properly			
	Door stops			
2. Windows	Adjusted properly			
	Weather-stripped			
	Washed			
3. Heating	Furnace			
	Heating system			
	Hot-water heater			
4. Interior Finish	Walls			
	Ceilings			
	Floors			
5. Painting	Interior			
	Exterior			
	Floors finished			
6. Electricity	Fixtures			
	Appliances			
	Outlets			
7. Plumbing	Toilets			
	Faucets			
	Drains			
8. Bathrooms	Fixtures			
	Mirrors and cabinets			
	Shower doors			
	Tile			
9. Trim	Base			
	Cabinets			
	Wardrobes			
10. Yard	Grading and landscaping			
	Walk, drives, stoops			
	Spouts, splash blocks			
11. Basement	Dry			
	Finished			
12. Utilities	Meters installed			
13. Other				

Customer_____ Construction Manager_____

INSPECTION PROCEDURE

FROM: General Manager

TO: Construction Manager

SUBJECT: Construction Inspection Record

Purpose

The purpose of this procedure is to establish a construction inspection record for documenting all inspections requested, made, approved, or rejected at the construction site each day by customers and by FHA, lender, and town inspectors.

Responsibility

The construction manager is responsible for the preparation and maintenance of construction inspection records.

Record Form

Attached is a sample of the form used to record daily inspections made at the construction site. The form provides for the recording of:

1. Project name and location
2. Manager of the project
3. Date of the record
4. Weather on that date
5. Inspections:
    a. FHA compliance—first, second, and third inspections
    b. Lender compliance—first through fourth inspections
    c. Town building department—first through fifth inspections
    d. Town health department:
        (1) Water inspection
        (2) Well inspection
        (3) Cesspool inspection
    e. Customer acceptance:
        (1) Closing inspection
        (2) Postclosing inspection
6. Job action:
    a. Job numbers for each inspection requested
    b. Job numbers for each inspection made
    c. Job numbers for each inspection approved
    d. Job numbers for each inspection rejected
7. Job comments—reasons for rejections, etc.

The construction inspection record is designed to serve as daily inspection report for the main office and also as a daily log for all inspections made.

Preparation Procedure
1. At the close of each workday the construction manager is to complete the construction inspection record in duplicate.
2. The original copy of the record is to be retained in the construction manager's field office construction inspection record file readily available for future reference and review with inspection agencies and main office personnel.
3. The carbon copy of the record is to be forwarded to the general manager for information on the status of inspections pending and made at the construction site.

APPROVED BY: _____

DATE: _____

Inspection Procedures 185

CONSTRUCTION INSPECTION RECORD

Project_____ Weather_____
Manager_____ Date_____

Inspections	Job Action		Approved	Rejected	Comments
	Requested	Made			
FHA Compliance					
1st					
2d					
3d					
Lender Compliance					
1st					
2d					
3d					
4th					
Town Building Dept.					
1st					
2d					
3d					
4th					
5th					
Town Health Dept.					
Well					
Water					
Cesspool					
Customer Acceptance					
Closing					
Postclosing					

CHAPTER 13

Marketing Procedures

MARKETING PROCEDURE

FROM: General Manager
TO: Marketing Manager
SUBJECT: Sales Option Agreements

Purpose
The purpose of this procedure is to establish a routine method for the preparation and processing of sales option agreements which serve as customer sales binders pending the receipt of their credit approvals and the signing of the preliminary sales agreement.

Responsibility
The marketing manager is responsible for supervising the preparation and processing of sales option agreements by sales personnel.

Sales Option Form
The company's standard sales option agreement form is to be used to bind customer sales until their credit has been approved and the preliminary sales agreement has been signed. If the credit report indicates that the customer

Marketing Procedures 187

is a poor credit risk, the deposit received is to be refunded and the sales option agreement canceled.

Procedure
1. Pending the receipt of a credit report and the preparation of the preliminary sales agreement, a sales option agreement is prepared by marketing and signed by customers upon receipt of their home deposits.
2. Three copies of the sales option agreement are to be signed and distributed as follows:
    a. One copy is given to the customer.
    b. One copy is retained by marketing in its sales option agreement file.
    c. The original copy with the customer's deposit is forwarded to the finance manager.
3. The finance manager retains the deposit, records the sales option, requests a credit report on the customer, and forwards the sales option agreement to the office secretary.
4. The office secretary opens a job folder for the customer in which the sales option agreement is filed.
5. Following the receipt of a favorable credit report on the customer, marketing prepares a preliminary sales agreement for signing with the customer. If the customer's report is unfavorable, the binder deposit is returned, the sales option agreement destroyed, and the customer's job folder closed.

APPROVED BY: _____

DATE: _____

MARKETING PROCEDURE

FROM: General Manager
TO: Marketing Manager
SUBJECT: Preliminary Sales Agreements

Purpose
The purpose of this procedure is to establish a routine method for the preparation and processing of preliminary sales agreements made between customers and the company.

Responsibility
The marketing manager is responsible for supervising the preparation and processing of preliminary sales agreements by sales personnel.

Sales Agreement Form
The company's standard preliminary sales agreement form is used for documenting all sales agreements entered with customers for their purchase of

homes. The preliminary sales agreement remains in force until it is superseded by the firm contract of sale.

Procedure
1. After the customer's credit status has been approved and the balance of the down payment received, the preliminary sales agreement is prepared for signature by the customer and the marketing manager.
2. After the preliminary sales agreement is signed, copies are distributed by marketing as follows:
   a. One copy is given to the customer.
   b. One copy is retained by marketing for its preliminary sales agreement file.
   c. The original copy with the customer's deposit is forwarded to the finance manager.
3. The finance manager retains the deposit, records the sale, and forwards the original copy of the preliminary sales agreement to the office secretary.
4. The office secretary files the preliminary sales agreement in the customer's job folder.
5. When the firm contract of sale is consummated with the customer, it is filed in the job folder with the preliminary sales agreement.

APPROVED BY: _____

DATE: _____

MARKETING PROCEDURE

FROM: General Manager
TO: Marketing Manager
SUBJECT: Contracts of Sale

Purpose
The purpose of this procedure is to establish a routine method for the preparation and processing of firm contracts of sale consummated by customers and the company for the sale of homes.

Responsibility
The marketing manager is responsible for supervising the preparation and processing of all contracts of sale.

Contract of Sale Form
The company's standard contract of sale form is to be used for documenting all firm sales contracts entered into with customers for their purchase of

Marketing Procedures 189

homes. The contract of sale remains in force until title is conveyed from the company to the customer at the sales closing.

Procedure
1. The marketing manager establishes the time and date for the signing of the contract of sale by the customer and the company.
2. The general manager and the customer meet at the company sales office and complete the sale by their signing of the contract of sale.
3. The marketing manager distributes six copies of the contract of sale as follows:
    a. The original copy is forwarded to the office secretary for file in the customer's job folder.
    b. One copy is given to the customer.
    c. Four copies are sent to the finance manager for use with mortgage commitment applications and title closings.

APPROVED BY: _____

DATE: _____

MARKETING PROCEDURE

FROM: General Manager
TO: Marketing Manager
SUBJECT: Customer Service Register

Purpose
The purpose of this procedure is to establish a means for recording and monitoring customer service activities administered by the marketing department.

Responsibility
The marketing manager is responsible for the preparation and maintenance of the customer service register.

Register Form
The customer service register consists of a three-ring binder with loose-leaf register forms for recording the descriptions of services and the dates they are provided for each customer's construction job. Customer service forms in the register are filed in a chronological sequence based on the dates that deposits were received for each job. Attached is a sample of the customer service register form.

Procedure
1. Upon receipt of a customer deposit, the marketing manager is to post the following information on a customer service form, after which the form is inserted in the customer service register:
   a. Customer's name
   b. Telephone number
   c. Address
   d. Project name
   e. Job number
   f. Date deposit received
2. As each of the subsequent customer service actions are completed, their transaction dates are posted in the column provided.
3. After the customer acceptance date has been recorded, the marketing manager is to remove the completed form and forward it to the office secretary for file in the customer's job folder.

APPROVED BY: _____

DATE: _____

Marketing Procedures

CUSTOMER SERVICE REGISTER

Customer_____ Telephone_____
Address_____
Project_____ Job No._____

Customer Services	Date
1. Customer deposit received	
2. Customer change orders received	
3. Sales agreement executed	
4. Sales agreement copy with letter of transmittal sent to customer	
5. Customer selections form with letter of transmittal sent to customer	
6. Customer selections completed	
7. Mortgage application made	
8. Mortgage application approved	
9. Report of title ordered	
10. Report of title completed	
11. Final survey ordered	
12. Final survey completed	
13. Closing date scheduled:	
a. Customer notified	
b. Customer attorney notified	
c. Title company notified	
d. Customer bank notified	
14. Utilities ordered:	
a. Gas	
b. Water	
c. Electric	
15. Deed prepared and approved	
16. Customer complaints resolved	
17. Customer acceptance of home	

Comments

MARKETING PROCEDURE

FROM: General Manager

TO: Marketing Manager

SUBJECT: Customer Selections Reports

Purpose
The purpose of this procedure is to establish a routine method for the processing of customer color, finish, and material selections.

Responsibility
The company's sales personnel are responsible for assisting customers with their color, finish, and material selections and for the preparation of customer selections reports.

Selection Form
Attached is a sample of the customer selections report form which must be completed for each customer. The critical stages of construction at which customer selections must be finalized are indicated on the report.

Procedure
1. When the foundation footings for the house have been installed, customer selections must be completed for paint colors, ceramic tile, bath fixtures, formica, kitchen cabinets, oven and range, range hood, and exterior siding.
2. Customer selections are entered on the report by the sales representative, dated, and endorsed with the customer's signature.
3. Four copies of the customer selections report are made and distributed as follows:
    a. One copy is given to the customer.
    b. One copy is forwarded to the construction manager.
    c. Two copies are retained by the sales representative.
4. The construction manager advises subcontractors of the customer selections through the use of job work orders and then forwards his selections report to the office secretary to file in the customer's job folder.
5. When the painting undercoat is applied, the customer's Kentile selections must be completed. The selections made are entered on both copies of the sales representative's selections report, dated, and signed by the customer.
6. The sales representative furnishes one copy of the completed selections report to the customer and forwards the other copy to the construction manager.
7. After issuing a job work order with the Kentile color selection to the resilient floor subcontractor, the construction manager forwards the selections report to the office secretary.

Marketing Procedures

8. The completed customer selections report is filed by the office secretary in the customer's job folder.

APPROVED BY: _____

DATE: _____

CUSTOMER SELECTIONS REPORT

Customer_____ Address_____
Project_____ Job No._____ Salesman_____

A. Paint Colors:	
Living Room	Family Room
Dining Room	Full Bath
Halls, Lower	Half Bath, Lower
Master Bedroom	Half Bath, Upper
Second Bedroom	Kitchen
Third Bedroom	Front Door
Den	Shutters, Trim
Halls, Upper	Shutters, Louvers
B. Ceramic Wall Tile:	C. Bath Fixtures:
Full Bath	Full Bath
Half Bath, Lower	Half Bath, Lower
Half Bath, Upper	Half Bath, Upper
D. Ceramic Floor Tile:	E. Formica:
Full Bath	Full Bath
Half Bath, Lower	Kitchen
Half Bath, Upper	F. Kitchen Cabinets
G. Oven and Range	H. Range Hood
I. Exterior Siding	
J. Kentile Floors: Kitchen	Family Room

Selection Approvals

Items A–I: When house footings poured	Customer	Date
Item J: When interior undercoat painted	Customer	Date

MARKETING PROCEDURE

FROM: General Manager
TO: Marketing Manager
SUBJECT: Weekly Sales Reports

Purpose

The purpose of this procedure is to establish a routine method for the preparation and distribution of weekly sales reports summarizing the sales activities of the marketing department.

Responsibility

The marketing manager is responsible for the preparation and distribution of weekly sales reports.

Report Form

A sample of the form used to summarize weekly sales activities is attached. Completion of the form is self-explanatory.

Procedure

1. The weekly sales report covering the previous week's sales activities is to be prepared on Monday mornings.
2. The marketing manager is to forward one copy of the report to the general manager and retain a duplicate copy in the marketing file.
3. The summary of sales activities listed on the report is to be reviewed jointly by the general manager and marketing manager each Monday afternoon.
4. Following the sales conference with the marketing manager, the general manager is to forward the original copy of the weekly sales report to the office secretary for the project file.

APPROVED BY: _____

DATE: _____

WEEKLY SALES REPORT

Project_____ Report No._____
Marketing Manager_____ Week Ending_____
Weekend Weather_____

Weekend Traffic_____

Sales Activities	This Week's Totals	Jobs Sold This Week				
Options Signed Pending						
Agreements Signed Pending						
Contracts Signed Pending						
Closings Held Scheduled						
Inventory Sold Not Sold						

Comments

_____
_____

Marketing Manager_____

CHAPTER 14

Filing Procedures

FILING PROCEDURE

FROM: General Manager
TO: Office Secretary
SUBJECT: Project File System

Purpose
The purpose of this procedure is to establish a routine method for the filing and control of all documents and correspondence which originate during the administrative life of the project.

Responsibility
The office secretary is responsible for the establishment and maintenance of a project file system for the retrieval and storage of all documents and correspondence relating to the overall administration of the project.

Project File Format
The organization structure of the project file system is to conform with the following outline:

Filing Procedures

A. *Project engineering files*
  1. Project plans and specifications
     a. Site drawings
     b. Site specifications
     c. Architectural drawings
     d. Architectural specifications
  2. Site analysis surveys
  3. Construction changes
     a. Company change orders
     b. Customer change orders
     c. Change order controls
  4. Architect correspondence

B. *Construction files*
  1. Subcontractors
     a. Subcontractor agreements
     b. Subcontractor amendments
     c. Subcontractor correspondence
     d. Subcontractor requests for quotations
     e. Subcontractor schedules
     f. Subcontractor register
  2. Vendors
     a. Purchase orders
     b. Purchase order register
     c. Vendor requests for quotations
     d. Vendor correspondence
     e. Vendor register
  3. Records and reports
     a. Daily field reports
     b. Daily field report register
     c. Customer complaint record
     d. Construction inspection record
     e. Construction status reports
     f. Construction variance reports

C. *Marketing files*
  1. Customer correspondence
  2. Customer service register
  3. Weekly sales reports

D. *Finance files*
  1. Construction mortgages
     a. Applications
     b. Commitments
     c. Location surveys
     d. Title reports
     e. Compliance inspection reports
     f. Property descriptions
  2. Overhead
     a. Personnel

b. Utilities
   c. Rentals
   d. Services
   e. Other
  3. Subcontractors
   a. Subcontractor agreements
   b. Subcontractor invoices
   c. Credit investigations
   d. Performance bonds
   e. Certificates of insurance
  4. Vendors
   a. Purchase orders
   b. Vendor invoices
 E. General correspondence files
  1. Customer correspondence
  2. Company correspondence
  3. Legal correspondence
 F. Job files
  1. Active job files
  2. Inactive job files
     (*Note:* See job file system procedure.)

Storage Files

All project correspondence and documents are to be retained in standard drawer files with the exception of the following:
1. Project drawings:
   a. Auto positives—stored in cubby files
   b. Working drawings—hung on rack files
2. Inactive drawings—stored in cubby files

Filing Procedure

1. Copies of all documents and correspondence initiated, processed, and received by the office secretary are to be filed in their respective storage folders or holders in accordance with the project file structure outlined in this procedure.
2. Individual procedures prescribe the correspondence and documents, and the number of each, which must be filed in the project file system.
3. Following the completion of all administrative activities on the project, the project file system is to be inactivated and transferred to record storage under the direction of the finance manager.

APPROVED BY:_____

DATE:_____

FILING PROCEDURE

FROM: General Manager
TO: Office Secretary
SUBJECT: Job File System

Purpose
The purpose of this procedure is to establish a routine method for the filing and control of all correspondence and documents which originate during the administrative life of each construction job.

Responsibility
The office secretary is responsible for the establishment and maintenance of a job file system for the retrieval and storage of all correspondence and documents relating to the administration of each construction job.

Job File Format
The organization structure of the filing system for each construction job (house) is to conform with the following outline:
A. Contracts folder
   1. Sales option agreement
   2. Preliminary sales agreement
   3. Contract of sale
B. Permits folder
   1. Building permit
   2. Fireplace permit
   3. Plumbing permit
   4. Certificate of occupancy
C. Mortgage folder
   1. Conditional commitment
   2. Firm commitment
   3. Plot plans
   4. Grading and location plans
D. Inspection folder
   1. Building department
   2. Health department
   3. Lender compliance
   4. Customer acceptance
E. Closing folder
   1. Final survey
   2. Final underwriter's certificate
   3. Title-closing statement
   4. Escrow agreement
F. Customer service folder
   1. Customer service form

    2. Customer selections report
    3. Customer change orders
    4. Customer complaints report
    5. Job work orders
G. *General correspondence folder*
    1. Customer correspondence
    2. Company correspondence

Storage Files
All job files are to be retained in standard cabinet drawer files.

Filing Procedure
1. Job files are to be activated by the office secretary upon receipt of a sales option agreement or preliminary sales agreement.
2. Copies of all job documents and correspondence which are subsequently initiated, processed, or received by the office secretary are filed in their respective job folders as afore-outlined in this procedure.
3. Individual procedures prescribe the correspondence and documents, and the number of each, which must be filed in their respective job files.
4. The job files are to be stored in their file drawers in the alphabetical sequence of customers' names.
5. The job file system is to be divided into two major sections:
    a. Active job files
    b. Inactive job files

After the closing of a sale, the job file is to be transferred by the office secretary from the active to the inactive section of the job file system.

                                      APPROVED BY: _____

                                                  DATE: _____

BIBLIOGRAPHY

"Air Force Systems Command," *Air Force Magazine*, September, 1961.

Aras, Restan M., and Julius Surkis: "PERT and CPM Techniques in Project Management," *Journal of the Construction Division*, Proceedings of the American Society of Civil Engineers, March, 1964.

Arrow Diagram Planning, Petroleum Chemicals Div., E. I. du Pont de Nemours & Co., Inc., Wilmington, Del., 1962.

Barnard, Chester I.: *The Functions of the Executive*, Harvard University Press, Cambridge, Mass., 1948.

Bittel, Lester R.: *Management by Exception*, McGraw-Hill Book Company, New York, 1964.

Buffa, Elwood S.: *Modern Production Management*, John Wiley & Sons, Inc., New York, 1963.

Bursk, Edward C., and John F. Chapman: *New Decision-making Tools for Managers*, Harvard University Press, Cambridge, Mass., 1963.

Chapin, Ned: *An Introduction to Automatic Computers*, D. Van Nostrand Company, Inc., Princeton, N.J., 1963.

"CPM: Builders Blueprint for Bigger Profits," *American Builder*, June, 1964.

CPM: A Plan for Progress, Technical Services Department, National Association of Home Builders, Washington, D.C.

Computers and Machines: Tools for Better Management Planning Control, The National Housing Center and the NAHB, *Journal of Homebuilding,* National Association of Home Builders, Washington, D.C., September, 1963.

"Critical Path Method: The New Way to Take Guesswork Out of Scheduling," *House and Home,* April, 1963.

Drucker, Peter F.: *The Practice of Management,* Harper & Row, Publishers, Incorporated, New York, 1954.

Ellis, David O., and Fred J. Ludwig: *Systems Philosophy,* Prentice-Hall, Inc., Englewood Cliffs, N.J., 1962.

Fisch, Gerald G.: *Organization for Profit,* McGraw-Hill Book Company, New York, 1964.

Fondahl, John W.: *A Non-computer Approach to the Critical Path Method for the Construction Industry,* Department of Civil Engineering, Stanford University, Stanford, Calif., 1962.

Frambes, Roland: "The Fact and the Fantasy of Management Systems," *Aerospace Management,* March, 1963.

Gregory, Robert H., and Richard L. Van Horn: *Business Data Processing and Programming,* Wadsworth Publishing Company, Inc., Belmont, Calif., 1963.

Horsinger, Vernon C.: "A Managers View of PERT/CPM," *Naval Engineers Journal,* April, 1966.

"How Computered CPM Saves Builders Money," *House and Home,* May, 1964.

Howard, Burl W.: "CPM: As Complete Project Management," *Journal of the Construction Division,* Proceedings of the American Society of Civil Engineers, May, 1965.

Hull, Seabrook: "The Systems Concept: Out of Adolescence and Here to Stay," *Missile Design and Development,* January, 1961.

Kushnerick, John P.: "The Zuckert-LeMay Management Team," *Aerospace Management,* December, 1961.

Learned, Edmund P., David N. Ulrich, and Donald R. Booz: *Executive Action,* Harvard University, Boston, 1951.

Livingston, J. Sterling: "Decision Making in Weapons Development," *Harvard Business Review,* Cambridge, Mass., January–February, 1958.

Livingston, J. Sterling, J. Ronald Fox, and Willard Fazar: "PERT Gains New Dimensions," *Aerospace Management,* January, 1962.

McCloskey, Joseph F., and Florence N. Trefethen: *Operations Research for Management,* The Johns Hopkins Press, Baltimore, 1954.

Mace, Myles L.: *The Growth and Development of Executives,* Harvard University, Boston, 1950.

Martino, R. L.: "How Useful Are Critical Path Methods?" *Chemical Engineering*, Sept. 14, 1964.
Miller, Robert W.: *Schedule, Cost, and Profit Control with PERT*, McGraw-Hill Book Company, New York, 1963.
Monaghan, John O.: "Do-It-Yourself CPM," *Construction Methods*, January, 1966.
Nelson, Richard L.: "A Managerial Look at the Science of Systems Analysis," *Naval Engineers Journal*, August, 1965.
O'Brien, James J.: *CPM in Construction Management*, McGraw-Hill Book Company, New York, 1965.
O'Brien, James J.: "CPM and PERT: Additions to Engineers' Vocabulary," *Naval Engineers Journal*, December, 1965.
PERT Guide for Management Use, PERT Coordinating Group: U.S. Department of Defense, Atomic Energy Commission, Bureau of the Budget, Federal Aviation Agency, National Aeronautics and Space Administration, Washington, D.C., 1963.
Peterson, Russell J.: "Critical Path Scheduling for Construction Jobs," August, 1962.
Polaris Management Fleet Ballistic Missile Program, Special Projects Office, U.S. Department of the Navy, Washington, D.C., September, 1962.
Reid, James C., Jr.: "The Systems Approach to Planning Advanced Ship Development," *Naval Engineers Journal*, August, 1966.
Sando, Francis A.: "CPM: What Factors Determine Its Success," *Architectural Record*, May, 1964.
Stonus, John J.: "Construction Preplanning Pays Off," *Civil Engineering*, November, 1964.
Systems Analysis in Decision-making, papers submitted at symposium sponsored by the Requirements Committee, Electronics Industries Association, at the Auditorium, Institute for Defense Analysis, Arlington, Va., June 23 and 24, 1966.
Systems Development and Management, Hearings before a Subcommittee on the Committee on Government Operations, U.S. House of Representatives, 87th Cong., 2d Sess., 1962, parts 1–2.
Systems Management-System Documentation, Air Force Regulation 375-4, U.S. Department of the Air Force, Washington, D.C., Jan. 23, 1961.
Systems Management-System Program Director, Air Force Regulation 375-3, U.S. Department of the Air Force, Washington, D.C., Jan. 23, 1961.
Van Krugel, E.: "Introduction to CPM," *Architectural Record*, September, 1964.

Waldron, A. James: *Fundamentals of Project Planning and Control,* A. James Waldron, Haddonfield, N.J., August, 1963.

Weapon/Support/Command and Control Systems Management, Air Force Regulation 375-1, U.S. Department of the Air Force, Washington, D.C., Jan. 23, 1961.

Weiser, Peter B.: "Operational Concepts and the Weapon System," *Astronautics,* December, 1960.

INDEX

Accounting and record keeping, 105
Activity, work, in work chart, 45–46
Activity arrows, 46
Activity plan, construction, semicustom
    home construction, 48–49
  work, plant labor, 56
Activity time estimates, construction,
    40–44
  job inspection, 44
  plant labor, 41
  subcontractor, 42–43
Administration, system approach, 113
Advertising by marketing manager, 109
Armed services and systems management
    approach, 13–16

Bar chart, 7–9
  home building use, 8
    limitations, 8
  manufacturing use, 7
Budget controls and finance manager,
    105
Building department inspections, town,
    173–174
Building permit applications, 134–135
Building permits, 39

Calendar, program control, 63, 64
Certificate of occupancy applications, 154
Cesspool inspections, 176
Change order, construction, 118–121
Change order control, 122–123
  form, 123
Changes and selections, and program
    control chart, 27
  in requirements planning, 38–39
  in resource plan, 67
Changes and selections lead-time sched-
    ule, 47, 50, 51, 67
Changes and selections list, 38
Closing inspections, 181
Commitment applications in finance pro-
    cedure, conditional, 128–129
  firm, 129–130
Complaints, customer, 147–150
Computer critical path method, 12
Computer programming techniques, 2,
    6
Computers for home building operations,
    18–20
Concurrent-flow scheduling, 62–64
Concurrent paths, 45–46
Construction, heavy, critical path method
    for, 10, 11, 13, 14

205

Index

Construction, in operations planning, 29
  and program control chart, 23
  (*See also* Home building)
Construction activity plan, 48–49
Construction activity starts directory, 65
Construction change orders, 118–121
  form, 123
Construction department and marketing manager, 109
Construction draw lead-time schedule, 55
Construction-flow network, 45
  concurrent path, 45
  critical path, 45
  event, 45
  slack arrows, 45
  slack time, 45
  work activity, 45
Construction-flow plan, 44–47, 56–57
  and critical path network, 44
  defined, 31
Construction inspection record, 183–185
  form, 185
Construction loan application, 124–125
Construction loan closings, 125
Construction loan receipts, 126
Construction loan receipts schedule, 127
Construction management techniques, 16
Construction manager, 96, 106–108
Construction manpower in requirements planning, 34–36
Construction materials, 37–38
Construction materials purchase list, 37
Construction operations, multi-site, 98–99
  single-site, 97, 98
Construction operations information, 23–27
  changes and selections, 27
  construction, 23
  inspection, 26
  plant labor, 27
  purchasing, 26
  subcontractors, 23
Construction operations system, 34
Construction performance, 74
Construction plan, 29
  defined, 40
Construction planning, 40–57
  construction-flow plan, 44
  construction-time estimates, 40
  resource-flow plan, 55
  resource lead time, 47
Construction procedures, 134–154
Construction program, balance and control, 80
  overall flow of, measuring, 74

Construction progress bars, 70
Construction schedule variance control, 77, 79
  form, 78
Construction schedule variance control reports, 145–146
  form, 146
Construction scheduling, 60–64
  defined, 60
  jobs, 61
  production, 61
  and production-flow control, 59–60
Construction starts, 63
Construction status evaluation, 69–74
Construction status report, 77, 78, 143–144
  form, 144
Construction systems management, 16–18
Construction-time estimates, 40–44
Construction variance control, 74–77, 84
Construction workday variance from schedule, 84–86
Construction workdays and job schedules, 61
Continuous-job home building operation, 59
Contracts of sale, 188–189
Control, delegation and, 100–110
Control chart, program (*see* Program control charts)
Cost estimates, 104
Crash operations, 80
Critical path, 45, 46
Critical path method (CPM), 8–14, 24–25
  acceptance in home building, 11–12
  computer service costs, 12
  distinguishing characteristics, 10–11
  as a project management tool, 11
  use in heavy construction industry, 10
Custom flow in operations scheduling, 58
Custom-order flow scheduling, 59
Custom-order home building, 59
Customer acceptance inspection, 180–182
Customer acceptance inspection report, 182
Customer complaint control, 149–150
  form, 150
Customer complaints, 147–148
  form, 148
Customer selection, 109
Customer selection reports, 192–193
  form, 193
Customer service register, 189–191
Customer services, 107

Index

Daily field report, 136–137
　form, 137
Daily field report register, 138–139
　form, 139
Delegation and control, 100–110
　authority, definition, 101
　delegation, definition, 101
Department managers, 93
Design by project engineer, 104
Drafting by project engineer, 104
du Pont de Nemours, E. I., 9

Efficiency, operations, 80
Electronic data processing techniques, 2, 18
　multi-programming, 18
Engineer, project, 104
Engineering procedures, 115–123
Engineering services, 103
Equipment, construction, 107
Escrow agreements, 132–133
Estimates, construction-time, 41–44
"Event," 45

Federal Housing Administration (see under FHA)
FHA compliance inspections, 177–178
FHA conditional commitment applications, 128–129
FHA firm commitment applications, 129–130
Field report, daily, 136–137
Field report register, daily, 138–139
File system, job, 199–200
　project, 196–199
Filing procedure, 196–200
Finance in operations planning, 29
Finance department, and construction manager, 108
　and marketing manager, 109
Finance manager, 96, 105–106
Finance procedures, 124–133
Financial plan, 29
Flow curves, program, 81–83
Flow plan, construction, 44–47, 56–57
Flow rates, job, 60
Flow scheduling, sequential, 62
Forecasting by marketing manager, 109
Functional management approach, 5
Functional organization chart, 93, 94
Functional organization structure, 95–97

Gantt, Henry L., 7
Gantt charts, 7–9, 24–25

General manager, 93, 95, 102–103
Growth, organization for, 90–99

Health department inspections, town, 174–176
Home building, continuous order, 59
　custom-order, 59
　industry, 2
　　economic structure, 2–3
　　expansion, 3
　multi-site, 98–99
　operating characteristics, 17
　organizations, 17
　physical aspects, 4
　restraints to growth of, 88
　semi-custom, 48–49, 62–64
　single-site, 97, 98
　in United States, 87
　after World War II, 2–6
Housing units, flow of, 60

Inspection procedures, 173–185
Inspections, job (see under Job inspection)
　and permits in requirements planning, 39–40
　and program control chart, 26–27
Integrated management systems, 15
Integrated operations systems, 17, 57

Job descriptions, 100–110
　as administrative tool, 100
　function performance requirements, 101
　for organization development and control, 101
Job file system, 199–200
Job flow rates in construction scheduling, 60
Job inspection in resource plan, 67
Job inspection activity time estimates, 44
Job inspections lead-time schedule, 54–55, 67
Job schedules, 61
Job status record, 70–73, 140–142
　form, 141
Job status section of program control chart, 74–77
Job work orders, 151–153
　form, 153

Labor, plant (see Manpower; Plant labor)

Land improvement, 103
Lead time, 47–55
　definition of, 47
Lead-time schedule, changes and selections, 50, 51
　construction draw, 55
　job inspection, 54, 55
　purchase order, 51, 52
　in resource scheduling, 65–67
　subcontractor, 51–54, 65
Lender compliance inspections, 178–180
Loan applications, 124–125
Loan closings, 125–126
Loan receipts, 126–128
Loans and finance manager, 105
Logic network technique, 9, 15, 16, 18

Management, in home building, 93, 95–96
　and job descriptions, 100–102
　management team, 101
　restraints, 88
　after World War II, 4–6
Management concepts, innovations in, 5
Management control techniques, 88–89, 103, 111–113
　relationship and dependencies, 112
　for successful home building growth, 89
Management controls, 103
　balance of, 17
Management scientists, 1–2
Management techniques, systems, 12–21
Managers, construction, 96, 106–108
　department, 93
　finance, 96, 105–106
　general, 93, 95, 102–103
　marketing, 95, 108–109
Manpower, construction, in requirements planning, 34–36
　(See also Plant labor)
Manufacturing industries, 2
　automation in, 4
　economic structures, 2
　expansion, 3
　inspection requirements, 4
Manufacturing and operations control techniques after World War II, 2–6
Market research and analysis, 109
Marketing department, 108
Marketing manager, 96, 108–109
Marketing operations planning, 28, 29
Marketing plan, 29
Marketing procedures, 186–195
Materials, construction, 37–38

Materials purchase list, construction, 37–38
Mortgage loans, 105
Multiprogramming, 18–19
　housing units, 19
　weapons programs, 18

Network symbols, 45

Office management, 105–106
Office secretary, 110
Operations, crash, 80
Operations control systems, 5
Operations control techniques, 1–27
　comparative analysis, 24–25
Operations controls, implementing, 68–86
Operations efficiency, 80
Operations flow, balance, 80
Operations flow system, 59
Operations information, construction, 23
Operations performance, deterrents to, 80
Operations planning, 1–6
　definition, 28
　developing, 28–57
　fundamentals of, 29–31
　general manager, 103
　home building, 3
　manufacturing, 3
　prerequisites for, 31–33
Operations research, 2, 3, 6
Operations restraints, 88
Operations schedules, 58–60
　establishing, 58–67
　general manager, 103
Operations scheduling, 58–60
　custom flow, 59
　definition, 58
　production flow, 59–60
Operations system, 5
Organization, development, and control, 87–89
　for growth, 90–99
　management methodology, 91
Organization chart, functional, 93, 94
Organization communication system, 101
Organization structures, defined, 100
　economic characteristics, 95
　flexibility for growth, 97
　functional elements, 93
　functional structures, 95–97
　for multi-site operations, 99
　total economic complex, 99

Index

Parallel-flow scheduling, 62
Permits and inspection list, 39
Permits and inspections in requirements planning, 39–40
PERT (program evaluation and review technique), 13, 15
Planning, by marketing manager, 109
　long range, homebuilding, 3
　manufacturing, 3
　operations (*see* Operations planning)
Planning goals defined, 30
Plans and specifications, architect, 33
　by project engineer, 104, 117–118
Plant labor, and program control chart, 27
　and resource plan, 65, 66
Plant labor activity time estimates, 40–44
Plant labor scheduling, 65
Plant labor work activity plan, 56
Plant labor work list, 34–35
Polaris Fleet Ballistic Missile Program, 13
Postclosing inspection, 181
Preliminary sales agreements, 187–188
President of organization, 93, 95, 102
　duties and responsibilities, 102
Procedures, description, 113
　systems and, 103, 111–114
Procurement and subcontracting, 107
Production control, defined, 70
Production control function, 68
Production control techniques, 68–77
Production flow, 59–60
　in operations scheduling, 58–60
　and program control chart, 71
Production-flow control, 60
Production-flow scheduling, 59
　concurrent-flow, 62
　parallel-flow, 62
　sequential-flow, 62
Production restraints, home building, 88
Production schedules, 62–64
Program control calendar, 63, 64
Program control charts, 21–27
　job status information, 74–77
Program control sheet, 22
Program control techniques, 77, 80–86
Program control work chart, 45
Program element progress bars, 20
Program elements, 20
Program evaluation and review technique (PERT), 13, 15
Program flow curves, 81–83
　concave line, 82
　convex line, 82
　inverted S-line, 83
　S-line, 83
　straight line, 81

Program network plan, 20
Program planning sheet, 21
Program progress curve, 20
Program schedules, 20
Program status evaluation, 80–84
Program variance control, 20–21, 84–86
Program variance control record, 84–86
Program variance control technique, 86
Program workdays, 20
Project engineer, 104
Project file system, 196–199
Project plans and specifications, 117–118
Purchase order lead-time schedule, 51, 52, 67
Purchase order register, 160–161
　form, 161
Purchase orders, 158–159
　form, 159
　in resource plan, 67
Purchasing procedure, 155–164
Purchasing and program control chart, 26
Purchasing system, 113

Quotation, requests for, 155–157

Record keeping, by finance manager, 105
Requests for quotation, 155–157
　form, 157
Requirements planning, 33–40
　changes and selections, 38
　construction manpower, 34
　construction materials, 37
　defined, 30
　permits and inspections, 39
Resource-flow plan, 40, 55–57
　defined, 31
Resource lead time, 47–55
　defined, 47
Resource plans, 29
Resource scheduling, 64–67
　changes and selections, 67
　defined, 64
　job inspections, 67
　plant labor, 65
　purchase orders, 67
　subcontractor trades, 64

Safety program, 107
Sale, contracts of, 188–189
Sales administration, 108
Sales agreements, preliminary, 187–188
Sales option agreements, 186–187
Sales planning and forecasting, 109

Sales promotion, 108
Sales reports, weekly, 194–195
Schedule performance and operations, 69
Secretary, office, 110
Selections (*see* Changes and selections)
Semi-custom home construction activity plan, 48, 49
Sequential-flow scheduling, 62
Site analysis survey, 104, 115–117
Site preparation, 135–136
Slack arrows, 45
Slack time, 45, 46
Specification list, standard, 30
Specifications and plans, 104, 117–118
Status reports (*see* Construction status report; Job status record)
Subcontracting, 107
Subcontractor and program control chart, 23, 26
Subcontractor activity time estimates, 42–43
Subcontractor agreements, 165–166
Subcontractor certifications, 170
Subcontractor lead-time schedule, 51–54, 65
Subcontractor register, 171–172
  form, 172
Subcontractor team, 35
Subcontractor trades, 64–65
Subcontractor work lists, 35–36
Subcontractor work schedule, 65, 66, 167–169
  form, 169
Supply contracts, 60
Symbols, network, 45
Systems and procedures, 103, 111–114
  definition, 112
Systems management, construction, 16–18
Systems management approach, 5, 13, 16–18
Systems management matrix, 18–21
  data processing simulation, 19
  operation input-output model, 21
Systems management techniques, 12–21
  of United States Air Force, 14

Time estimates, construction, 40–44
  job inspection activity, 44
  plant labor activity, 41
  subcontractor activity, 42–43
Time-network analysis, 11
Title closing, 131–132

Town building department inspections, 39, 173–174
Town health department inspections, 174–176
Trades, start-up, finishing, and structural, 82–83

United States Air Force, 13
United States Navy, 13
Univac Applications Research Section, 9

Variance control, construction, 74–77, 84
  program, 20–21, 84–86
Variance control record, program, 84–86
Variance control reports, construction schedule, 145–146
Vendor certifications, 162
Vendor register, 163–164
  form, 164

Water lateral inspections, 175
Weapons systems, 13–14, 16
  American Armed Services, 12
  management concepts, 6, 12, 13, 15
  multi-programming, 18
  organizations, 17
  and PERT, 13
  program control, 15
  programs, 13, 15–16
  Soviet Union, 13
  technology, 13
Weekly sales reports, 194–195
  form, 195
Well inspections, 175–176
Work activity, in job status record, 72–73
  in work chart, 45–46
Work activity plan, plant labor, 56
Work chart, 45, 61
Work lists, 34–36
  plant labor, 34
  subcontractor, 35–36
Work orders, job, 151–153
Work schedule, subcontractor, 65, 66
Workday variance from schedule, construction, 84–86
Workdays, in activity time estimates, 41–44
  in resource lead time, 47–55
  (*See also* Job status record)